STRUCTURE ELUCIDATION BY NMR IN ORGANIC CHEMISTRY

STRUCTURE ELUCIDATION BY NMR IN ORGANIC CHEMISTRY
A Practical Guide

Eberhard Breitmaier
University of Bonn, Germany

Translated by **Julia Wade**

JOHN WILEY & SONS
Chichester · New York · Brisbane · Toronto · Singapore

Other Wiley Editorial Offices

John Wiley & Sons, Inc., 605 Third Avenue,
New York, NY 10158-0012, USA

Jacaranda Wiley Ltd, G.P.O. Box 859, Brisbane,
Queensland 4001, Australia

John Wiley & Sons (Canada) Ltd, 22 Worcester Road,
Rexdale, Ontario M9W 1L1, Canada

John Wiley & Sons (SEA) Pte Ltd, 37 Jalan Pemimpin #05-04,
Block B, Union Industrial Building, Singapore 2057

Library of Congress Cataloging-in-Publication Data

Breitmaier, E.
    [Vom NMR-Spektrum zur Strukturformel organischer Verbindungen. English]
    Structure elucidation by NMR in organic chemistry / E. Breitmaier ;
    translated by Julia Wade.
        p.   cm.
    Includes bibliographical references and index.
    ISBN 0 471 93745 2 (cloth) ISBN 0 471 93381 3 (paper)
    1. Organic compounds—Structure.   2. Nuclear magnetic
    resonance spectroscopy.   I. Title.
    QD476.B6713   1993
    547.3′ 0877—dc20

92-18531
CIP

British Library Cataloguing in Publication Data

A catalogue record for this book is available from the British Library

ISBN 0 471 93745 2 (cloth)
ISBN 0 471 93381 3 (paper)

Printed in Great Britain by Bookcraft (Bath) Limited

CONTENTS

PREFACE

These days, virtually all students of chemistry, biochemistry, pharmacy and related subjects learn how to deduce molecular structures from nuclear magnetic resonance (NMR) spectra. Undergraduate examinations routinely set problems using NMR data, and masters' and doctoral theses describing novel synthetic or natural products provide many examples of how powerful NMR has become in structure elucidation. Existing texts on NMR spectroscopy generally deal with the physical background of the newer and older techniques as well as the relationships between NMR parameters and chemical structures. Few, however, convey the know-how of structure determination using NMR, namely the strategy, techniques and methodology by which molecular structures are deduced from NMR spectra.

This book, based on many lectures and seminars, attempts to provide advanced undergraduates and graduate students with a systematic, readable and inexpensive introduction to the methods of structure determination by NMR. It starts with a deliberately concise survey of the basic terms, parameters and techniques dealt with in detail in other books, which this workbook is not intended to replace. An introduction to basic strategies and tactics of structure elucidation using one- and two-dimensional NMR methods then follows in Chapter 2. Here, the emphasis is always on how spectra and associated parameters can be used to identify structural fragments. This chapter does not set out to explain the areas usually covered, such as the basic principles of NMR, pulse sequences and theoretical aspects of chemical shift and spin–spin coupling. Instead, it presents those topics that are essential for the identification of compounds or for solving structures, including the atom connectivities, relative configuration and conformation, absolute configuration, intra- and intermolecular interactions and, in some cases, molecular dynamics. Following the principle of 'learning by doing,' Chapter 3 presents a series of case studies, providing spectroscopic details for 50 compounds that illustrate typical applications of NMR techniques in the structural characterisation of both synthetic and natural products. The level of difficulty, the sophistication of the techniques and the methodology required increase from question to question, so that all readers will be able to find material suited to their knowledge and ability. One can work independently, solve the problem from the spectra and check the result in the formula index, or follow the detailed solutions given in Chapter 4. The spectroscopic details are presented in a way that makes the maximum possible information available at a glance, requiring minimal page-turning. Chemical shifts and coupling constants do not have to be read off from scales but are presented numerically, allowing the reader to concentrate directly on problem solving without the need for tedious routine work.

My thanks must go especially to the Deutsche Forschungsgemeinschaft and to the Federal State of Nordrhein Westfalia for supplying the NMR spectrometers, and to Dr S. Sepulveda-Boza (Heidelberg), Dr K. Welmar (Bonn), Professor R. Negrete (Santiago,

Chile), Professor B. K. Cassels (Santiago, Chile), Professor Chen Wei-Shin (Chengdu, China), Dr A. M. El-Sayed and Dr A. Shah (Riyadh, Saudi Arabia), Professor E. Graf and Dr M. Alexa (Tübingen), Dr H. C. Jha (Bonn), Professor K. A. Kovar (Tübingen) and Professor E. Röder and Dr A. Badzies-Crombach (Bonn) for contributing interesting samples to this book. Also, many thanks are due to Dr P. Spuhler and to the publishers for their endeavours to meet the demand of producing a reasonably priced book.

Bonn, Autumn 1989   E. Breitmaier
        Autumn 1991

The cover shows the ^{13}C NMR spectrum of α- and β-D-xylopyranose at mutarotational equilibrium (35% α: 65% β, in deuterium oxide, 100 MHz, ^1H broadband decoupling) with the INADEQUATE contour plot. An interpretation of the plot according to principles described in Section 2.2.7 gives the CC bonds of the two isomers and confirms the assignment of the signals in Table 2.12.

SYMBOLS AND ABBREVIATIONS

APT: Attached Proton Test, a modification of the J-modulated spin-echo experiment to determine CH multiplicities, a less sensitive alternative to DEPT

CH COLOC: Correlation via long-range coupling, detects CH connectivities through two or three bonds in the CH COSY format

COSY: Correlated spectroscopy, two-dimensional shift correlations

CH COSY: Correlation via one-bond CH coupling, also referred to as HETCOR (heteronuclear shift correlation), provides carbon-13 and proton chemical shifts of CH bonds as cross signals in a δ_C *versus* δ_H diagram, assigns all CH bonds of the sample

HH COSY: Correlation via HH coupling with square symmetry because of equal shift scales in both dimensions provides all detectable HH connectivities of the sample

DEPT: Distortionless enhancement by polarisation transfer, differentiation between CH, CH_2 *and* CH_3 by positive or negative signal amplitudes, using improved sensitivity of polarisation transfer

CW: Continuous wave or frequency sweep, the older and less sensitive basic technique of NMR detection

FID: Free induction decay, decay of the induction (transverse magnetisation) back to equilibrium, following excitation of a nuclear spin by a radiofrequency pulse, in a way which is free from the influence of the radiofrequency field; this signal is Fourier-transformed to the FT NMR spectrum

FT NMR: Fourier transform NMR, the newer and more sensitive, less time consuming basic technique of NMR detection

INADEQUATE: Incredible natural abundance double quantum transfer experiment, segregates AB or AX systems due to one-bond CC couplings, detects CC bonds present in the sample (Carbon skeleton)

HMBC: Heteronuclear multiple bond correlation, inverse CH correlation via long-range coupling, same format as (^{13}C detected) CH COLOC but more sensitive because of ^1H detection

HMQC: Heteronuclear multiple quantum coherence, inverse CH correlation via one-bond carbon-proton coupling, same format as (^{13}C detected) CH COSY but much more sensitive because of ^1H detection

NOE: Nuclear Overhauser effect, change of signal intensities during decoupling experiments

NOESY: Nuclear Overhauser effect spectroscopy, detection of NOE in an HH COSY analogue format, traces out closely spaced protons in larger molecules

ROESY: Rotating frame NOESY, modified NOESY with suppressed spin diffusion, detects closely spaced protons in smaller molecules

TOCSY: Total correlation spectroscopy in the *HH* COSY format, detects proton-proton connectivities in addition to the usual proton-proton couplings (^2J, ^3J, ^4J, ^5J) as detected by *HH* COSY

J or ^1J: nuclear spin-spin coupling constant (in Hz) through a single bond (one-bond coupling)

^2J, ^3J: nuclear spin-spin coupling constant (in Hz) through two or three bonds (*geminal* and *vicinal* coupling)

Multiplet abbreviations:

S, s: singlet
D, d :   doublet
T, t :   triplet
Q, q :   quartet
Qui, qui :   quintet
Sxt, sxt :   sextet
Sep, sep :   septet
o :   overlapping
b :   broad

Capital letters: multiplets which are the result of coupling through one bond

Lower-case letters : multiplets which are the result of coupling through several bonds

δ : Contrary to IUPAC conventions, chemical shifts δ in this book are given in ppm, thus enabling the reader to differentiate at all times between chemical shift values (ppm) and coupling constants (Hz); ppm (parts per million) is in this case the ratio of two frequencies of different orders of magnitude, Hz/MHz, 1 : 10^6.

Italicised data and multiplet abbreviations refer to ^1H

1 SHORT INTRODUCTION TO BASIC PRINCIPLES AND METHODS

1.1 Chemical shift[1-3]

Chemical shift relates the Larmor frequency of a nuclear spin to its chemical environment. The Larmor frequency is the precession frequency of a nuclear spin in a static magnetic field (Fig. 1.1).

Because the Larmor frequency is proportional to the strength of the magnetic field, there is no absolute scale of chemical shift. Thus, a frequency difference (Hz) is measured from the resonance of a standard substance [tetramethylsilane (TMS) in ^1H and ^{13}C NMR] and divided by the absolute value of the Larmor frequency of the standard (several MHz), which itself is proportional to the strength of the magnetic field. The chemical shift is therefore given in parts per million (ppm, δ scale), because a frequency difference in Hz is divided by a frequency in MHz, these values being in a proportion of $1:10^6$.

Chemical shift is principally caused by the electrons in the molecule having a *shielding*

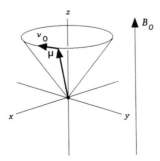

Fig. 1.1. Nuclear precession: nuclear charge and nuclear spin give rise to a magnetic moment of nuclei such as protons and carbon-13. The vector μ of the magnetic moment precesses in a static magnetic field with the Larmor frequency v_0 about the direction of the magnetic flux density vector B_0

effect on the nuclear spin. More precisely, the electrons cause a *shielding field* which opposes the external magnetic field: the precession frequency of the nuclear spin (and in turn the size of its chemical shift) is therefore reduced. An atomic nucleus (e.g. a proton) whose shift is reduced is said to be *shielded* (high shielding field); an atom whose shift is increased is said to be *deshielded* (low shielding field) (Fig. 1.2).

1.2   Spin–spin coupling[1-3]

Indirect or *scalar coupling* of nuclear spins through covalent bonds causes the splitting of NMR signals into multiplets in high-resolution NMR spectroscopy in the solution state. The *direct* or *dipolar coupling* between nuclear spins through space is only observed for solid-state NMR. In solution such coupling is cancelled out by molecular motion.

1.3   Coupling constants[1-3]

The *coupling constant* is the frequency difference J in Hz between two multiplet lines. Unlike chemical shift, the frequency value of a coupling constant does not depend on the strength of the magnetic field. In high-resolution NMR a distinction is made between coupling through one bond (^1J or simply J, *one-bond* couplings) and coupling through

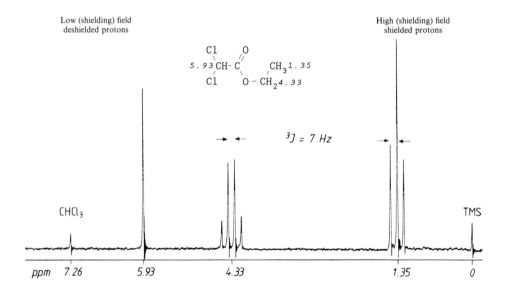

Fig. 1.2. ^1H NMR spectrum of ethyl dichloroacetate (CDCl$_3$, 25 °C, 60 MHz). The proton of the C*H*Cl$_2$ group is less shielded (more strongly deshielded) in comparison with the protons of the C*H*$_2$ and C*H*$_3$ residues

several bonds, e.g. two bonds (^2J, *geminal* couplings), three bonds (^3J, *vicinal* couplings) or four or five bonds (^4J and ^5J, long-range couplings).

For example, the CH_2 and CH_3 protons of the ethyl group in Fig. 1.2 are separated by three bonds; their (*vicinal*) coupling constant is $^3J = 7\,Hz$.

1.4   Signal multiplicity (multiplets)[1-3]

The *signal multiplicity* is the extent to which an NMR signal is split as a result of spin–spin coupling. Signals which show no splitting are known as *singlets* (*s*). Those with two, three, four, five, six or seven lines are known as *doublets* (*d*), *triplets* (*t*), *quartets* (*q*, Figs 1.2 and 1.3). *quintets* (*qui*), *sextets* (*sxt*) and *septets* (*sep*), respectively, but only where the lines of the multiplet signal are of equal distance apart, and the one coupling constant is therefore shared by them all. Where two or three different coupling constants produce a multiplet, this is referred to as a two- or three-fold multiplet, respectively, e.g. a *doublet of doublets* (*dd*, Fig. 1.3), or a *doublet of doublets of doublets* (*ddd*, Fig. 1.3). If both coupling constants of a doublet of doublets are sufficiently similar ($J_1 \approx J_2$), the middle signals overlap, thus generating a '*pseudotriplet*' ('*t*', Fig. 1.3).

The ^1H NMR spectrum of ethyl dichloroacetate (Fig. 1.2), as an example, displays a triplet for the CH_3 group (two *vicinal H*), a quartet for the OCH_2 group (three *vicinal H*) and a singlet for the $CHCl_2$ fragment (no *vicinal H* for coupling).

1.5   Spectra of first and higher order[2, 3]

First-order spectra (*multiplets*) are observed where the coupling constant is small compared with the frequency difference of chemical shifts between the coupling nuclei. This is referred to as an A_mX_n spin system, where nucleus A has the smaller and nucleus X has the considerably larger chemical shift. An AX system (Fig. 1.4) consists of an A doublet and an X doublet with the common coupling constant J_{AX}.

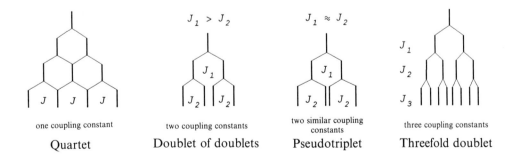

Fig. 1.3.   Quartet, doublet of doublets, pseudotriplet and threefold doublet (doublet of doublets of doublets)

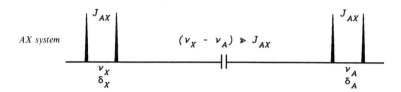

Fig. 1.4. Two-spin system of type AX with a chemical shift difference which is large compared with the coupling constants (schematic)

Multiplicity rules apply for first-order spectra (A_mX_n systems): When coupled, n nuclei of an element X with nuclear spin quantum number $I_x = \frac{1}{2}$ produce a splitting of the A signal into $n + 1$ lines; the relative intensities of the individual lines of a first-order multiplet are given by the coefficients of the Pascal triangle (Fig. 1.5).

The protons of the ethyl group of ethyl dichloroacetate (Fig. 1.2) as examples give rise to an A_3X_2 system with the coupling constant $^3J_{AX} = 7\,Hz$; the A protons (with smaller shift) are split into a triplet (two *vicinal* protons X, $n_X + 1 = 3$); the X protons appear as a quartet because of three *vicinal* A protons ($n_A + 1 = 4$).

For a given number, n, of coupled nuclear spins of spin quantum number I_x, the A signal will be split into ($2nI_x + 1$) multiplet lines (e.g. Fig. 1.9).

Spectra of higher order multiplicity occur for systems where the coupling constant is of similar magnitude to the chemical shift difference between the coupled nuclei. Such a case is referred to as an A_mB_n system, where nucleus A has the smaller and nucleus B the larger chemical shift.

An AB system (Fig. 1.6) may consist, for example, of an A doublet and a B doublet with the common coupling constant J_{AB}, where the external signal of both doublets is attenuated and the internal signal is enhanced. This is referred to as an AB *effect*, a 'roofing' symmetric to the centre of the AB system. 'Roofing' is frequently observed in proton NMR spectra, even in practically first order spectra (Fig. 1.2, ethyl quartet and triplet).

$n=0$	Singlet						1					
1	Doublet					1	:	1				
2	Triplet				1	:	2	:	1			
3	Quartet			1	:	3	:	3	:	1		
4	Quintet		1	:	4	:	6	:	4	:	1	
5	Sextet	1	:	5	:	10	:	10	:	5	:	1
6	Septet	1	:	6	:	15	:	20	:	15	:	6	:	1

Fig. 1.5. Relative intensities of first-order multiplets (Pascal triangle)

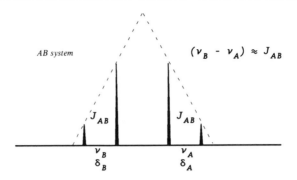

Fig. 1.6. Two-spin system of type *AB* with a small chemical shift difference compared to the coupling constant (schematic)

1.6 Chemical and magnetic equivalence[2, 3]

Chemical equivalence: atomic nuclei in the same chemical environment are chemically equivalent and thus show the same chemical shift. The 2,2'- and 3,3'-protons of a 1,4-disubstituted benzene ring, for example, are chemically equivalent because of molecular symmetry.

Magnetic equivalence: chemically equivalent nuclei are magnetically equivalent if they display the same coupling constants with all other nuclear spins of the molecule. For example, the 2,2'-(*AA'*) and 3,3'-(*X, X'*) protons of a 1,4-disubstituted benzene ring such as 4-nitroanisole are not magnetically equivalent, because the 2-proton *A* shows an *ortho* coupling with the 3-proton *X* (*ca 7–8 Hz*), but displays a different *para* coupling with the 3'-proton *X'* (*ca 0.5–1 Hz*). This is therefore referred to a an *AA'XX'* system rather than an A_2X_2 system (e.g. Fig. 2.6).

para coupling: $^5J_{AX'} = 0.5-1\ Hz$

ortho coupling: $^3J_{AX} = 7-8\ Hz$

4-Nitroanisole

1.7 Continuous wave (CW) and Fourier transform (FT) NMR spectra[2–6]

There are two basic techniques for recording high-resolution NMR spectra.

In the older *CW technique*, the frequency or field appropriate for the chemical shift range of the nucleus (usually ^1H) is swept by a continuously increasing (or decreasing)

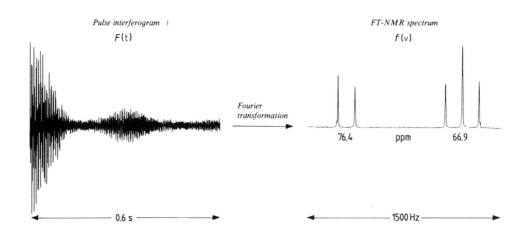

Fig. 1.7. Pulse interferogram and FT ^{13}C NMR spectrum of glycerol, $(HOCH_2)_2CHOH$, in D_2O at 25 °C and 100 MHz

radio-frequency. The duration of the sweep is long, typically 2 Hz/s, or 500 s for a sweep of 1000 Hz, corresponding to 10 ppm in 100 MHz proton NMR spectra. This monochromatic excitation therefore takes a long time to record. Figure 1.2, for example, is a CW spectrum.

In the *FT technique*, the whole of the Larmor frequency range of the observed nucleus is excited by a radiofrequency pulse. This causes transverse magnetisation to build up in the sample. Once excitation stops, the transverse magnetisation decays exponentially with the time constant T_2 of spin–spin relaxation provided the field is perfectly homogeneous. In the case of a one-spin system, the corresponding NMR signal is observed as an exponentially decaying alternative voltage (*free induction decay*, FID); multi-spin systems produce an interference of several exponentially decaying alternating voltages, the *pulse interferogram* (Fig. 1.7). The frequency of each alternating voltage is the difference between the individual Larmor frequency of one specific kind of nucleus and the frequency of the exciting pulse. The Fourier transformation (FT) of the pulse interferogram produces the Larmor frequency spectrum; this is the *FT NMR spectrum* of the type of nucleus being observed. Fourier transformation of the interferogram is performed with the help of a computer with a calculation time of less than 1 s.

The main advantage of the *FT* technique is the short time required for the procedure (about 1 s per interferogram). Within a short time a large number of individual interferograms can be accumulated, thus averaging out electronic noise (*FID accumulation*), and making the *FT* method the preferred approach for less sensitive NMR probes involving isotopes of low natural abundance (^{13}C, ^{15}N). Almost all of the spectra in this book are *FT* NMR spectra.

1.8 Spin decoupling[2, 3, 5, 6]

Spin decoupling (double resonance) is an NMR technique in which, to take the simplest example, an AX system, the splitting of the A signal due to J_{AX} coupling is removed if the sample is irradiated by a second radiofrequency which resonates with the Larmor frequency of the X nucleus. The A signal then appears as a singlet; at the position of the X signal interference is observed between the X Larmor frequency and the decoupling frequency. If the A and X nuclei are the same isotope (e.g. protons), this is referred to as *selective homonuclear decoupling*. If A and X are different, e.g. carbon-13 and protons, then it is referred to as *heteronuclear decoupling*.

Figure 1.8 illustrates homonuclear decoupling experiments with the CH protons of 3-aminoacrolein. These give rise to an AMX system (Fig. 1.8a). Decoupling of the aldehyde proton X (Fig. 1.8b) simplifies the NMR spectrum to an AM system ($^3J_{AM} = 12.5\,Hz$); decoupling of the M proton (Fig. 1.8c) simplifies to an AX system ($^3J_{AX} = 9\,Hz$). These experiments reveal the connectivities of the protons within the molecule.

In ^{13}C NMR spectroscopy, three kinds of heteronuclear spin decoupling are used.

In *proton broadband decoupling* of ^{13}C NMR spectra, decoupling is carried out unselectively across a frequency range which encompasses the whole range of the proton shifts. The spectrum then displays up to n singlet signals for the n non-equivalent C atoms of the molecule.

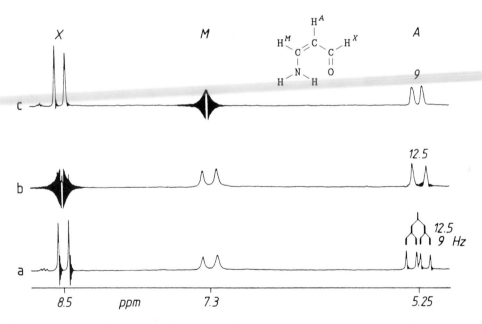

Fig. 1.8. Homonuclear decoupling of the CH protons of 3-aminoacrolein (CD$_3$OD, 25 °C, 90 MHz). (a) ^1H NMR spectrum; (b) decoupling at 8.5 ppm; (c) decoupling at 7.3 ppm. At the position of the decoupled signal in (b) and (c) interference beats are observed because of the superposition of the two very similar frequencies

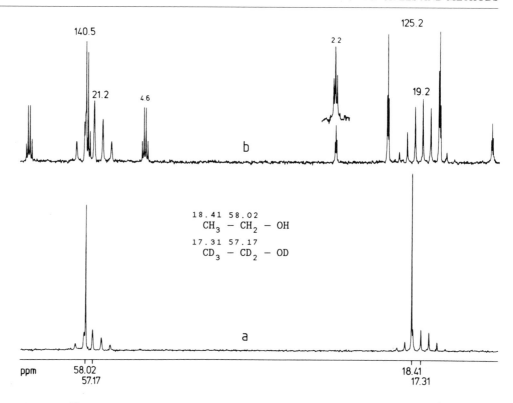

Fig. 1.9. ^{13}C NMR spectra of a mixture of ethanol and hexadeuterioethanol (27:75 v/v, 25 °C, 20 MHz). (a) ^{1}H broadband decoupled; (b) without decoupling. The deuterium isotope effect $\delta_{CH} - \delta_{CD}$ on ^{13}C chemical shifts is 1.1 and 0.85 ppm for methyl and methylene carbon nuclei, respectively

Figure 1.9 demonstrates the effect of proton broadband decoupling in the ^{13}C NMR spectrum of a mixture of ethanol and hexadeuterioethanol. The CH_3 and CH_2 signals of ethanol appear as intense singlets upon ^{1}H broadband decoupling while the CD_3 and CD_2 resonances of the deuterated compound still display their septet and quintet fine structure; deuterium nuclei are not affected by ^{1}H decoupling because their Larmor frequencies are far removed from those of protons; further, the nuclear spin quantum number of deuterium is $I = 1$; in keeping with the general multiplicity rule ($2nI_X + 1$, Section 1.5), triplets, quintets and septets are observed for CD, CD_2 and CD_3 groups, respectively.

In *selective proton decoupling* of ^{13}C NMR spectra, decoupling is performed at the precession frequency of a specific proton. As a result, a singlet only is observed for the attached C atom. *Off-resonance* conditions apply to the other C atoms. For these the individual lines of the CH multiplets move closer together, and the relative intensities of the multiplet lines change from those given by the Pascal triangle; external signals are attenuated whereas internal signals are enhanced. *Selective ^{1}H decoupling* of ^{13}C NMR spectra was used for assignment of the CH connectivities (CH bonds) before the much more efficient CH COSY technique (see Section 2.2.8) became routine. *Off-resonance*

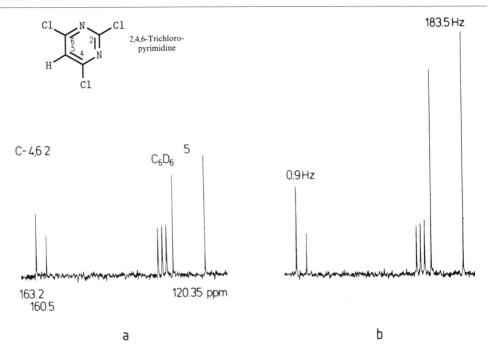

Fig. 1.10. ^{13}C NMR spectra of 2,4,6-trichloropyrimidine (C_6D_6, 75% v/v 25 °C, 20 MHz). (a) ^{13}C NMR spectrum without proton decoupling; (b) NOE enhanced coupled ^{13}C NMR spectrum (gated decoupling)

decoupling of the protons was helpful in determining CH multiplicities before better methods became available (see Section 2.2.3).

In *pulsed* or *gated decoupling* of protons (broadband decoupling only *between* FIDs), coupled ^{13}C NMR spectra are obtained in which the CH multiplets are enhanced by the *nuclear Overhauser effect* (NOE, see Section 1.9). This method is used when CH coupling constants are required for structure analysis because it enhances the multiplets of carbon nuclei attached to protons; the signals of quaternary carbons two bonds apart from a proton are also significantly enhanced. Figure 1.10 demonstrates this for the carbon nuclei in the 4,6-positions of 2,4,6-trichloropyrimidine.

Quantitative analysis of mixtures is achieved by evaluating the integral steps of ^1H NMR spectra. This is demonstrated in Fig. 1.11a for 2,4-pentanedione which occurs as an equilibrium mixture of 87% enol and 13% diketone. A similar evaluation of the ^{13}C integrals in ^1H broadband decoupled ^{13}C NMR spectra fails in most cases because signal intensities are influenced by nuclear Overhauser enhancements and relaxation times and these are usually specific for each individual carbon nucleus within a molecule. As a result, deviations are large (81–93% enol) if the keto–enol equilibrium of 2,4-pentane-dione is analysed by means of the integrals in the ^1H broadband decoupled ^{13}C NMR spectrum (Fig. 1.11b). *Inverse gated decoupling*, involving proton broadband decoupling only during the FIDs, helps to solve the problem. This technique provides ^1H broadband decoupled ^{13}C NMR spectra with suppressed nuclear Overhauser effect so that signal

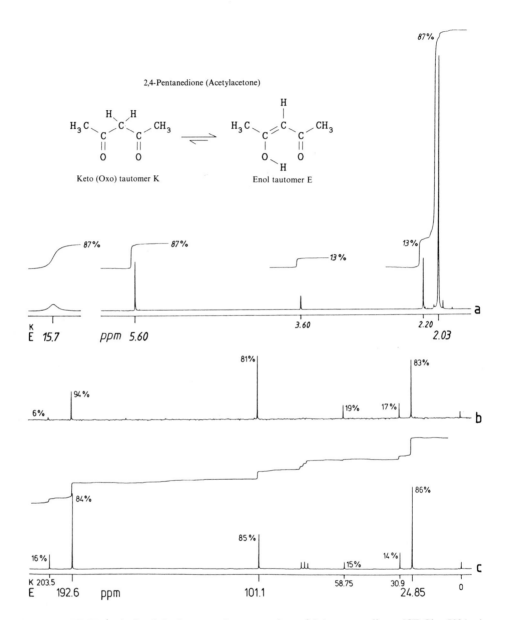

Fig. 1.11. NMR analysis of the keto–enol tautomerism of 2,4-pentanedione (CDCl$_3$, 50% v/v, 25 °C, 60 MHz for ^1H, 20 MHz for ^{13}C). (a) ^1H NMR spectrum with integrals (result: keto:enol = 13:87); (b) ^1H broadband decoupled ^{13}C NMR spectrum; (c) ^{13}C NMR spectrum obtained by inverse gated ^1H decoupling with integrals [result: keto:enol = 15:85 ($\pm$1)]

intensities can be compared and keto-enol tautomerism of 2,4-pentanedione, for example, is analysed more precisely as shown in Fig. 1.11c.

1.9 Nuclear Overhauser effect[2, 3]

The *nuclear Overhauser effect* (NOE, also an abbreviation for nuclear Overhauser enhancement) causes the change in intensity (increase or decrease) during decoupling experiments. The maximum possible NOE in high-resolution NMR of solutions depends on the gyromagnetic ratio of the coupled nuclei. Thus, in the homonuclear case such as proton-proton coupling, the NOE is much less than 0.5, whereas in the most frequently used heteronuclear example, proton decoupling of ^{13}C NMR spectra, it may reach 1.988. Instead of the expected signal intensity of 1, the net result is to increase the signal intensity threefold $(1 + 2)$. In proton broadband and gated decoupling of ^{13}C NMR spectra, NOE enhancement of signals by a factor of as much as 2 is routine, as was shown in Figs 1.9 and 1.10.

1.10 Relaxation, relaxation times[3, 6]

Relaxation refers to all processes which regenerate the Boltzmann distribution of nuclear spins on their precession states and the resulting equilibrium magnetisation along the static magnetic field. Relaxation also destroys the transverse magnetisation arising from phase coherence of nuclear spins built up upon NMR excitation.

Spin-lattice relaxation is the steady (exponential) build-up or regeneration of the Boltzmann distribution (equilibrium magnetisation) of nuclear spins in the static magnetic field. The lattice is the molecular environment of the nuclear spin with which energy is exchanged.

The *spin-lattice relaxation time*, T_1, is the time constant for spin-lattice relaxation which is specific for every nuclear spin. In FT NMR spectroscopy the spin-lattice relaxation must 'keep pace' with the exciting pulses. If the sequence of pulses is too rapid, e.g. faster than $3T_{1\ max}$ of the 'slowest' C atom of a molecule in carbon-13 resonance, a decrease in signal intensity is observed for the 'slow' C atom due to the spin-lattice relaxation getting 'out of step.' For this reason, quaternary C atoms can be recognised in carbon-13 NMR spectra by their weak signals.

Spin-spin relaxation is the steady decay of transverse magnetisation (phase coherence of nuclear spins) produced by the NMR excitation where there is perfect homogeneity of the magnetic field. It is evident in the shape of the FID (*free induction decay*), as the exponential decay to zero of the transverse magnetisation produced in the pulsed NMR experiment. The Fourier transformation of the FID signal (time domain) gives the FT NMR spectrum (frequency domain).

The *spin-spin relaxation time*, T_2, is the time constant for spin-spin relaxation which is also specific for every nuclear spin (approximately the time constant of FID). For small- to medium-sized molecules in solution $T_2 \approx T_1$. The value of T_2 of a nucleus determines the width of the appropriate NMR signal at half-height ('half-width') according to the uncertainty relationship. The smaller is T_2, the broader is the signal. The more rapid is the molecular motion, the larger are the values of T_1 and T_2 and the sharper are the signals ('*motional narrowing*'). This rule applies to small- and medium-sized molecules of the type most common in organic chemistry.

Chemical shifts and coupling constants reveal the static structure of a molecule; relaxation times reflect molecular dynamics.

2 RECOGNITION OF STRUCTURAL FRAGMENTS BY NMR

INTRODUCTION TO TACTICS AND STRATEGIES OF STRUCTURE ELUCIDATION BY ONE- AND TWO-DIMENSIONAL NMR

2.1 Functional groups

2.2.1 ^1H CHEMICAL SHIFTS

Many functional groups can be identified conclusively by their ^1H chemical shifts (Table 2.1).[1-3] Important examples are listed in Table 2.1, where the ranges for the proton shifts are shown in decreasing sequence: aldehydes (*9.5–10.5 ppm*), acetals (*4.5–6 ppm*), alkoxy (*4–5.5 ppm*) and methoxy functions (*3.5–4 ppm*), N-methyl groups (*3–3.5 ppm*) and methyl residues attached to double bonds such as C=C or C=X (X = N, O, S) or to aromatic and heteroaromatic skeletons (*1.8–2.5 ppm*).

Small shift values for CH or CH_2 protons may indicate cyclopropane units. Proton shifts distinguish between alkyne CH (generally *2.5–3.2 ppm*), alkene CH (generally *4.5–6 ppm*) and aromatic/heteroaromatic CH (*6–9.5 ppm*), and also between π-electron-rich (pyrrole, furan, thiophene, *6–7 ppm*) and π-electron-deficient heteroaromatic compounds (pyridine, *7.5–9.5 ppm*).

2.1.2 DEUTERIUM EXCHANGE

Protons which are bonded to heteroatoms (XH protons, X = O, N, S) can be identified in the ^1H NMR spectrum by using deuterium exchange (treatment of the sample with a small amount of D_2O or CD_3OD). After the deuterium exchange:

$$RXH + D_2O \rightleftharpoons RXD + HDO$$

the XH proton signals in the ^1H NMR spectrum disappear. Instead, the HDO signal appears at approximately *4.8 ppm*. Those protons which can be identified by D_2O

Table 2.1. ^1H chemical shift ranges for organic compounds

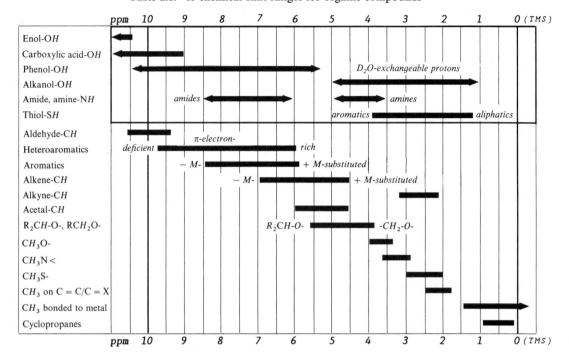

exchange are indicated as such in Table 2.1. As a result of D_2O exchange, XH protons are often not detected in the ^1H NMR spectrum if this is obtained using a deuteriated protic solvent (e.g. CD_3OD).

2.1.3 ^{13}C CHEMICAL SHIFTS

The ^{13}C chemical shift ranges for organic compounds[1, 4–6] in Table 2.2 show that many carbon-containing functional groups can be identified by the characteristic shift values in the ^{13}C NMR spectra.

For example, various carbonyl groups have distinctive shifts. Ketonic carbonyl functions appear as singlets falling between 190 and 220 ppm, with cyclopentanone showing the largest shift; although aldehyde signals between 185 and 205 ppm overlap with the shift range of keto carbonyls, they appear in the coupled ^{13}C NMR spectrum as doublet CH signals. Quinone carbonyl occurs between 180 and 190 ppm while the carbonyl C atoms of carboxylic acids and their derivatives are generally found between 160 and 180 ppm. However, the ^{13}C signals of phenoxy carbon atoms, carbonates, ureas (carbonic acid derivatives), oximes and other imines also lie at about 160 ppm so that additional information such as the empirical formula may be helpful for structure elucidation.

Other functional groups that are easily differentiated are cyanide (110–120 ppm) from isocyanide (135–150 ppm), thiocyanate (110–120 ppm) from isothiocyanate (125–140 ppm), cyanate (105–120 ppm) from isocyanate (120–135 ppm) and aliphatic C atoms which are bonded to different heteroatoms or substituents (Table 2.2). Thus ether-methoxy generally appears between 55 and 62 ppm, ester-methoxy at 52 ppm; N-methyl

Table 2.2. ¹³C chemical shift ranges for organic compounds

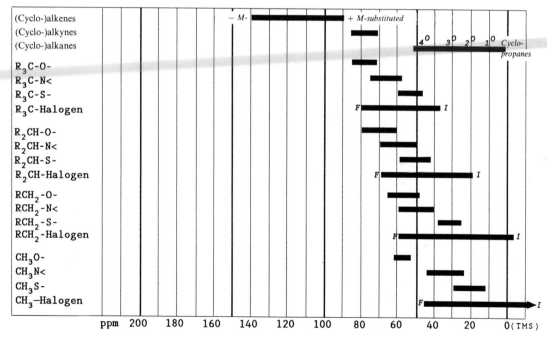

generally lies between 30 and 45 ppm and S-methyl at about 25 ppm. However, methyl signals at 20 ppm may also arise from methyl groups attached to C=X or C=C double bonds, e.g. as in acetyl, CH_3—CO—.

If an H atom in an alkane R—H is replaced by a substituent X, the ^{13}C chemical shift δ_C in the α-position increases proportionally to the electronegativity of X ($- I$ effect). In the β-position, δ_C generally also increases, whereas it decreases at the C atom γ to the substituent (γ-effect, see Section 2.3.4). More remote carbon atoms remain almost uninfluenced ($\Delta\delta_C \approx 0$).

In contrast to ^1H shifts, ^{13}C shifts cannot in general be used to distinguish between aromatic and heteroaromatic compounds on the one hand and alkenes on the other (Table 2.2). Cyclopropane carbon atoms stand out, however, by showing particularly small shifts in both the ^{13}C and the ^1H NMR spectra. By analogy with their proton resonances, the ^{13}C chemical shifts of π electron-deficient heteroaromatics (pyridine type) are larger than those of π electron-rich heteroaromatic rings (pyrrole type).

Substituent effects (substituent increments) tabulated in more detail in the literature[1-6] demonstrate that ^{13}C chemical shifts of individual carbon nuclei in alkenes and aromatic and heteroaromatic compounds can be predicted approximately by means of mesomeric effects (resonance effects). Thus, an electron donor substituent D [D = OCH_3, SCH_3, $N(CH_3)_2$] attached to a C=C double bond shields the β-C atom and the β-proton ($+ M$ effect, smaller shift), whereas the α-position is deshielded (larger shift) as a result of substituent electronegativity ($- I$ effect).

The reversed polarity of the double bond is induced by a π electron-accepting substituent A (A = C=O, C≡N, NO_2): the carbon and proton in the β-position are deshielded ($- M$ effect, larger shifts).

These substituents have analogous effects on the C atoms of aromatic and heteroaromatic rings. An electron donor D (see above) attached to the benzene ring deshields the (substituted) α-C atom ($- I$ effect). In contrast, in the ortho and para positions (or comparable positions in heteroaromatic rings) it causes a shielding ($+ M$ effect, smaller ^1H and ^{13}C shifts), whereas the meta positions remain almost unaffected.

+ M effect or (π-electron)-donor effect: shielding in the o- and o'- and p-positions

$$\delta_H < 7.26 \text{ ppm}; \quad \delta_C < 128.5 \text{ ppm}$$

An electron-accepting substituent A (see above) induces the reverse deshielding in *ortho* and *para* positions ($+ M$ effect, larger ^1H and ^{13}C shifts), again with no significant effect on *meta* positions.

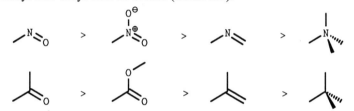

— M effect or (π-electron)-donor effect: deshielding in the o- and o'- and p-positions

$$\delta_H > 7.26 \text{ ppm}; \quad \delta_C > 128.5 \text{ ppm}$$

2.1.4 ^{15}N CHEMICAL SHIFTS

Frequently the ^{15}N chemical shifts[7-9] (Table 2.3) of molecular fragments and functional groups containing nitrogen complement their ^1H and ^{13}C shifts. The ammonia scale[7] of ^{15}N shifts used in Table 2.3 shows very obvious parallels with the TMS scale of ^{13}C shifts. Thus, the ^{15}N shifts (Table 2.3) decrease in size in the sequence nitroso, nitro, imino, amino, following the corresponding behaviour of the ^{13}C shifts of carbonyl, carboxy, alkenyl and alkyl carbon atoms (Table 2.2).

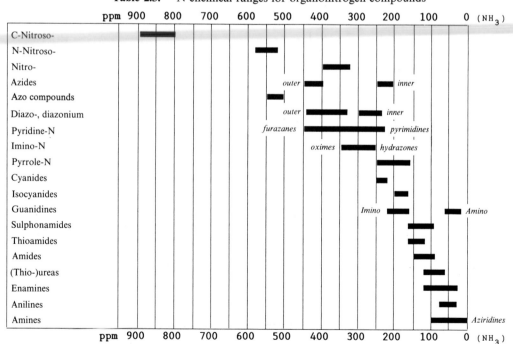

Table 2.3. ^{15}N chemical ranges for organonitrogen compounds

ppm	900	800	700	600	500	400	300	200	100	0 (NH$_3$)
C-Nitroso-		▬								
N-Nitroso-				▬						
Nitro-						▬				
Azides					*outer* ▬			*inner* ▬		
Azo compounds					▬					
Diazo-, diazonium				*outer* ▬		▬		*inner* ▬		
Pyridine-N					*furazanes* ▬	▬		*pyrimidines* ▬		
Imino-N						*oximes* ▬		*hydrazones* ▬		
Pyrrole-N								▬		
Cyanides								▬		
Isocyanides								▬		
Guanidines							*Imino* ▬		*Amino* ▬	
Sulphonamides								▬		
Thioamides								▬		
Amides								▬		
(Thio-)ureas									▬	
Enamines									▬	
Anilines									▬	
Amines									▬	*Aziridines* ▬
ppm	900	800	700	600	500	400	300	200	100	0 (NH$_3$)

The shift decrease found in ^{13}C NMR spectra in the sequence

$$\delta_{\text{alkenes, aromatics}} > \delta_{\text{alkynes}} > \delta_{\text{alkanes}} > \delta_{\text{cyclopropanes}}$$

also applies to the analogous N-containing functional groups, ring systems and partial structures (Tables 2.2 and 2.3):

$$\delta_{\text{imines, pyridines}} > \delta_{\text{nitriles}} > \delta_{\text{amines}} > \delta_{\text{aziridines}}$$

2.2 Skeletal structure (atom connectivities)

2.2.1 *HH* MULTIPLICITIES

The splitting (signal multiplicity) of ^1H resonances often reveals the spatial proximity of the protons involved. Thus it is possible to identify structural units such as those which often occur in organic molecules simply from the appearance of multiplet systems and by using the $n + 1$ rule.

The simplest example is the AX or AB system for a $—CH^A—CH^{X(B)}—$unit; Fig. 2.1 shows the three typical examples: (a) the AX system, with a large shift difference between the coupled protons H^A and H^X; (b) the AB system, with a small difference in the coupled nuclei (H^A and H^B) relative to the coupling constant J_{AB}, and (c) the AB system, with a very small shift difference [$(v_B - v_A) \leqslant J_{AB}$] verging on the A_2 case, whereby the outer

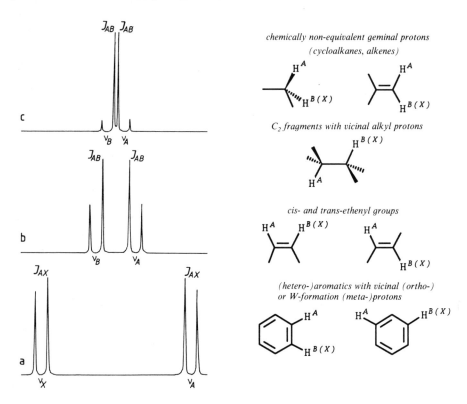

Fig. 2.1. *AX(AB)* systems and typical molecular fragments

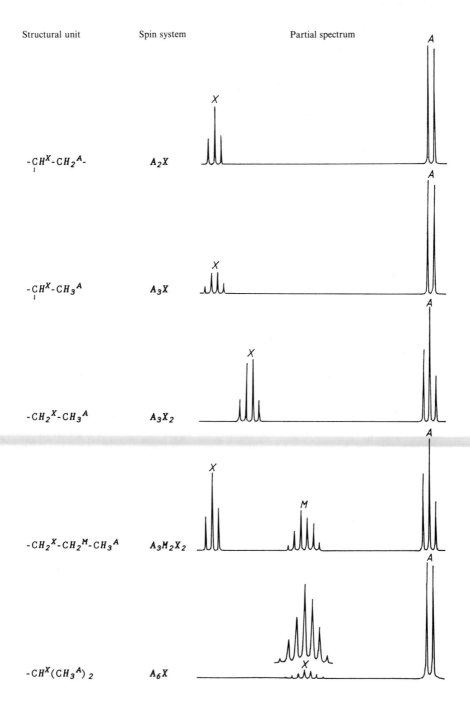

Structural unit   Spin system   Partial spectrum

$-CH^X-CH_2^A-$ A_2X

$-CH^X-CH_3^A$ A_3X

$-CH_2^X-CH_3^A$ A_3X_2

$-CH_2^X-CH_2^M-CH_3^A$ $A_3M_2X_2$

$-CH^X(CH_3^A)_2$ A_6X

Fig. 2.2. Easy to recognise A_mX_n systems and their typical molecular fragments

signals are very strongly suppressed by the strong roofing effect (*AB* effect). Figure 2.2 shows the *¹H* NMR partial spectra of a few more structural units which can easily be identified.

Structure elucidation does not necessarily require the complete analysis of all multiplets in complicated spectra. If the coupling constants are known, the characteristic fine structure of the single multiplet almost always leads to identification of a molecular fragment and, in the case of alkenes and aromatic or heteroaromatic compounds it may even lead to the elucidation of the complete substitution pattern.

2.2.2   CH MULTIPLICITIES

The multiplicities of ^{13}C signals due to $^{1}J_{CH}$ coupling (splitting occurs due to *CH* coupling across one bond) indicates the bonding mode of the C atoms, whether quaternary (R_4C, singlet S), tertiary (R_3CH, doublet D), secondary (R_2CH_2, triplet T) or

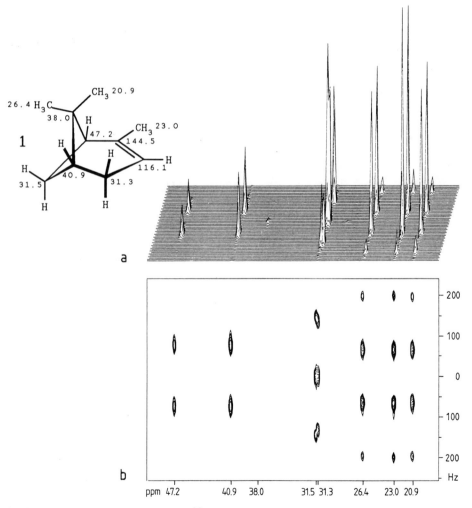

Fig. 2.3.   *J*-resolved two-dimensional ^{13}C NMR spectra of α-pinene (**1**) [in $(CD_3)_2CO$, 25 °C, 50 MHz]. (a) Stacked plot; (b) contour plot

primary (RCH_3, quartet Q). Coupled ^{13}C NMR spectra which have been enhanced by NOE are particularly suitable for indicating CH multiplicities (gated decoupling).[5,6] Where the sequence of signals in the spectra is too dense, evaluation of spin multiplicities may be hampered by overlapping. In the past this has been avoided by compression of the multiplet signals using off-resonance decoupling[5,6] of the protons. More modern techniques are the J-modulated spin–echo technique (attached proton test, APT)[10,11] and J-resolved two-dimensional ^{13}C NMR spectroscopy,[12,13] which use J-modulation and the DEPT sequence.[14,15] Figure 2.3, shows a series of J-resolved ^{13}C NMR spectra of α-pinene (**1**) as a contour plot and as a stacked plot. The purpose of the experiment is apparent; ^{13}C shift and J_{CH} coupling constants are shown in two frequency dimensions so that signal overlaps occur less often.

The J-modulated spin–echo technique[10,11] and the DEPT technique[14,15] are pulse sequences, which transform the information of the CH signal multiplicity and of spin–spin coupling into phase relationships (positive and negative amplitudes) of the ^{13}C signals in the proton decoupled ^{13}C NMR spectra. The DEPT technique benefits from a ^{1}H–^{13}C polarisation transfer which increases the sensitivity by up to a factor of 4. For this reason, this technique provides the quickest way of determining the ^{13}C–^{1}H multiplicities. Figure 2.4 illustrates the application of the DEPT technique to the analysis of the CH multiplets of α-pinene (**1**). Routinely the result will be the subspectrum (b) of all CH carbon atoms in addition to a further subspectrum (c), in which, besides the CH carbon atoms, the CH_3 carbon atoms also show positive amplitude, whereas the CH_2

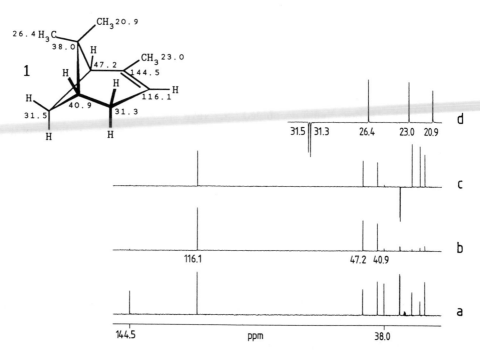

Fig. 2.4. CH multiplicities of α-pinene (**1**) (hexadeuterioacetone, 50 MHz). (a) ^{1}H broadband decoupled ^{13}C NMR spectrum; (b) DEPT subspectrum of CH; (c) DEPT subspectrum of all C atoms which are bonded to H (CH and CH_3 positive, CH_2 negative); (d) an expansion of a section of (c). Signals from two quaternary C atoms, three CH units, two CH_2 units and three CH_3 units can be seen

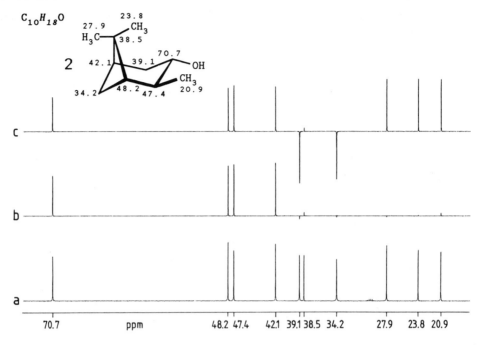

Fig. 2.5.   *CH* multiplicities of isopinocampheol (**2**), $C_{10}H_{18}O$ [$(CD_3)_2CO$, 25 °C, 50 MHz]. (a) ^1H broadband decoupled ^{13}C NMR spectrum; (b) DEPT *CH* subspectrum; (c) DEPT subspectrum of all C atoms which are bonded to H(*CH* and and *CH₃* positive, *CH₂* negative)

carbon atoms appear as negative. Quaternary C atoms do not appear in the DEPT subspectra; accordingly, they may be identified as the signals which appear additionally in the ^1H broadband decoupled ^{13}C NMR spectra.

Figure 2.4 illustrates the usefulness of *CH* multiplicities for the purpose of structure elucidation. The addition of all C, *CH*, *CH₂* and *CH₃* units leads to a part formula C_xH_y,

$$2C + 3CH + 2CH_2 + 3CH_3 = C_2 + C_3H_3 + C_2H_4 + C_3H_9 = C_{10}H_{16}$$

which contains all of the *H* atoms which are bonded to C. Hence the result is the formula of the hydrocarbon part of the molecule, e.g. that of α-pinene (**1**) (Fig. 2.4).

If the *CH* balance given by the *CH* multiplicities differs from the number of *H* atoms in the molecular formula, then the additional *H* atoms are bonded to heteroatoms. The ^{13}C NMR spectrum in Fig. 2.5 shows, for example, for isopinocampheol (**2**), $C_{10}H_{18}O$, a quaternary C atom (C), four *CH* units (C_4H_4), two *CH₂* units (C_2H_4) and three *CH₃* groups (C_3H_9). In the *CH* balance, $C_{10}H_{17}$, one *H* is missing when compared with the molecular formula, $C_{10}H_{18}O$; the compound therefore contains an *OH* group.

2.2.3 *HH* COUPLING CONSTANTS

Since spin–spin coupling[2,3] through bonds occurs because of the interaction between the magnetic moments of the atomic nuclei and the bonding electrons, the coupling constants[2,3] reflect the bonding environments of the coupled nucei. In ^1H NMR

Table 2.4. HH coupling constants (Hz) of some typical units in alicyclic, alkene and alkyne units[2, 3]

spectroscopy *geminal* coupling through *two* bonds ($^2J_{HH}$) and *vicinal* coupling through *three* bonds ($^3J_{HH}$) provide insight into the nature of these bonds.

Geminal HH coupling, $^2J_{HH}$, depends characteristically on the polarity and hybridisation of the C atom on the coupling path and also on the substituents and on the HCH bond angle. Thus $^2J_{HH}$ coupling can be used to differentiate between a cyclohexane ($-12.5\ Hz$), a cyclopropane ($-4.5\ Hz$) or an alkene ($2.5\ Hz$), and to show whether electronegative heteroatoms are bonded to methyl groups (Table 2.4). In cyclohexane and norbornane derivatives the w-shaped arrangement of the bonds between protons attached to alternate C atoms leads to distinctive $^4J_{HH}$ coupling (w-couplings, Table 2.4).

Vicinal HH coupling constants, $^3J_{HH}$, are especially useful in determining the *relative configuration* (see Section 2.3.1). However, they also reflect a number of other distinguishing characteristics, e.g. the ring size for cycloalkenes (a low value for small rings) and the α-position of electronegative heteroatoms. The latter are remarkable for their small coupling constants $^3J_{HH}$ (Table 2.5). The heteroaromatics furan, thiophene, pyrrole and pyridine can be distinguished because of the characteristic effects of the electronegative heteroatoms on their $^3J_{HH}$ couplings (Table 2.5).

The coupling constants of *ortho* ($^3J_{HH} = 7\ Hz$), *meta* ($^4J_{HH} = 1.5\ Hz$) and *para* protons ($^5J_{HH} \leqslant 1\ Hz$) in benzene and naphthalene ring systems are especially useful in structure elucidation (Table 2.5). With naphthalene and other condensed (hetero-) aromatics, knowledge of 'zig zag' coupling ($^5J_{HH} = 0.8\ Hz$) is helpful in deducing substitution patterns.

The *HH* coupling constants of pyridine (Table 2.5) reflect the positions of the coupling protons relative to the nitrogen ring. There is a particularly clear difference here between the protons in the 2- and 3-positions ($^3J_{HH} = 5.5\ Hz$) and those in the 3- and 4-positions ($^3J_{HH} = 7.6\ Hz$). Similarly, *HH* coupling constants in five-membered heteroaromatic rings, in particular the $^3J_{HH}$ coupling of the protons in the 2- and 3-positions, allow the

Table 2.5. *HH coupling constants (Hz) of aromatic and heteroaromatic compounds*[2, 3]

$^3J_{HH}$	$^4J_{HH}$	$^5J_{HH}$
7.5	1.5	0.7
8.3	1.3	0.7
7.0	0.7	0.8

Pyridines:

5.5	7.6	1.9	1.6	0.9

0.4

Five-membered heterocycles ($X = O$, NH, S):

1.8	3.4	0.9	1.5
2.6	3.5	1.3	2.1
4.8	3.5	1.0	2.8

$X = O$
NH
S

heteroatoms to be identified (the more electronegative the heteroatom, the smaller is the value of $^3J_{HH}$).

In the case of alkenes and aromatic and heteroaromatic compounds, analysis of a single multiplet will often clarify the complete substitution pattern. A few examples will illustrate the procedure.

If, for example, four signals are found in regions appropriate for benzene ring protons (*6.5–8.5 ppm*, four protons on the basis of the height of the integrals), then the sample is a disubstituted benzene (Fig. 2.6). The most effective approach is to analyse a multiplet with a clear fine structure and as many coupling constants as possible, e.g. consider the

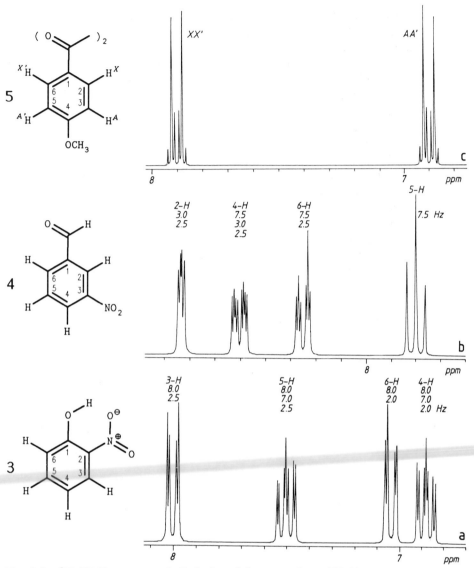

Fig. 2.6. ^{1}H NMR spectra of disubstituted benzene rings (CDCl$_3$, 25 °C, 200 MHz). (a) *o*-Nitrophenol (**3**); (b) *m*-nitrobenzaldehyde (**4**); (c) 4,4'-dimethoxybenzil (**5**)

threefold doublet at *7.5 ppm* (Fig. 2.6a); it shows two *ortho* couplings (*8.0* and *7.0 Hz*) and one *meta* coupling (*2.5 Hz*); hence relative to the *H* atom with a shift value of *7.5 ppm*, there are two protons in *ortho* positions and one in a *meta* position; hence the molecule must be an *ortho*-disubstituted benzene (*o*-nitrophenol, **3**).

A *meta* disubstituted benzene (Fig. 2.6 b) shows only two *ortho* couplings ($^{3}J_{HH} = 7.5\ Hz$) for one signal (*7.8 ppm*) whereas another signal (*8.74 ppm*) exhibits only two *meta* couplings ($^{4}J_{HH} = 3.0$ and *2.5 Hz*). In both cases one observes either a triplet (*7.8 ppm*) or a doublet of doublets (*8.74 ppm*) depending on whether the couplings ($^{3}J_{HH}$ or $^{4}J_{HH}$) are equal or different.

The $AA'XX'$ systems[2,3] which are normally easily recognisable from their symmetry identify *para*-disubstituted benzenes such as 4,4'-dimethoxybenzil (5) or 4-substituted pyridines.

This method of focusing on a ^1H multiplet of clear fine structure and revealing as many HH coupling constants as possible affords the substitution pattern for an alkene or an aromatic or a heteroaromatic compound quickly and conclusively. One further principle normally indicates the *geminal*, *vicinal* and *w* relationships of the protons of a molecule, the so-called HH connectivities, i.e. that *coupled nuclei have identical coupling constants*. Accordingly, once the coupling constants of a multiplet have all been established, the appearance of one of these couplings in another multiplet identifies (and assigns) the coupling partner. This procedure, which also leads to the solutions to problems 1–10, may be illustrated by means of two typical examples.

In Fig. 2.7 the ^1H signal with a typical aromatic shift of *7.1 ppm* shows a doublet of doublets with *J*-values of *8.5 Hz* (*ortho* coupling, $^3J_{HH}$) and *2.5 Hz* (*meta* coupling, $^4J_{HH}$). The ring proton in question therefore has two protons as coupling partners, one in the *ortho* position (*8.5 Hz*) and another in the *meta* position (*2.5 Hz*), and moreover these are in such an arrangement as to make a second *ortho* coupling impossible. Thus the benzene ring is 1,2,4-trisubstituted (6). The ring protons form an AMX system, and in order to compare frequency dispersion and 'roofing' effects this is shown first at 100 MHz and then also at 200 MHz. The *para* coupling $^5J_{AX}$, which is less frequently visible, is also resolved. From the splitting of the signal at *7.1 ppm* (H^M) a 1,2,3-trisubstituted benzene

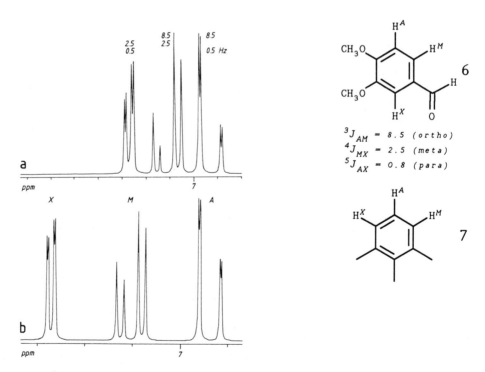

Fig. 2.7. ^1H NMR spectrum of 3,4-dimethoxybenzaldehyde (6) [aromatic shift range, CDCl₃, 25 °C, (a) 100 MHz (b) 200 MHz]

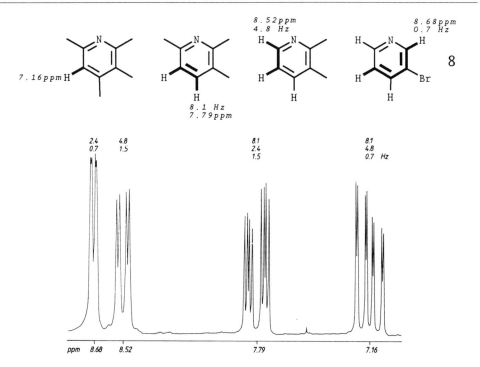

Fig. 2.8. ^1H NMR spectrum of 3-bromopyridine (**8**) (CDCl$_3$, 25 °C, 90 MHz)

ring (**7**) might have been considered. In this case, however, the *ortho* proton (H^A) would have shown a second *ortho* coupling to the third proton (H^X).

The application of the principle that coupled nuclei will have the same coupling constant enables the ^1H NMR spectrum to be assigned completely (Fig. 2.7). The *ortho* coupling, $^3J_{AM} = 8.5\ Hz$, is repeated at *6.93 ppm* and allows the assignment of H^A; the *meta* coupling, $^4J_{MX} = 2.5\ Hz$, which appears again at *7.28 ppm*, gives the assignment of H^X.

The four signals in the ^1H NMR spectrum of a pyridine derivative (Fig. 2.8) show first that it is a monosubstituted derivative. The signal with the smallest shift (*7.16 ppm*) splits into a threefold doublet with coupling constants *8.1, 4.8* and *0.7 Hz*. The two $^3J_{HH}$ couplings of *8.1* and *4.8 Hz* belong to a β proton of the pyridine ring according to Table 2.5. Step by step assignment of all three couplings (Fig. 2.8) unequivocally leads to a pyridine ring **8** substituted in the 3-position. Again, signals are assigned following the principle that coupled nuclei will have the same coupling constant; the coupling constants identified from Table 2.5 for the proton at *7.16 ppm* are then sought in the other multiplets.

2.2.4   CH COUPLING CONSTANTS

One-bond CH coupling constants J_{CH} ($^1J_{CH}$) are proportional to the s character of the hybrid bonding orbitals of the coupling carbon atom (Table 2.6):

$J_{CH} = 500 \times s$, where $s = 0.25, 0.33$ and 0.5

for sp^3-, sp^3- and sp-hybridised C atoms, respectively.

With the help of these facts, it is possible to distinguish between alkyl-C ($J_{CH} \approx$ 125 Hz), alkenyl- and aryl-C ($J_{CH} \approx 165$ Hz) and alkynyl-C ($J_{CH} \approx 250$ Hz), e.g. as in problem 13.

It is also useful for structure elucidation that J_{CH} increases with the electronegativity of the heteroatom or substituent bound to the coupled carbon atom (Table 2.6).

From typical values for J_{CH} coupling, Table 2.6 shows:

In the chemical shift range for aliphatic compounds

cyclopropane rings (*ca* 160 Hz);

oxirane (epoxide) rings (*ca* 175 Hz);

cyclobutane rings (*ca* 135 Hz);

O-alkyl groups (145–150 Hz);

N-alkyl groups (140 Hz);

acetal-C atoms (*ca* 170 Hz at 100 ppm);

terminal ethynyl groups (*ca* 250 Hz).

In the chemical shift range for alkenes and aromatic and heteroaromatic compounds

enol ether fragments (furan, pyrone, isoflavone, 195–200 Hz);

2-unsubstituted pyridine and pyrrole (*ca* 180 Hz);

2-unsubstituted imidazole and pyrimidine (> 200 Hz).

Geminal CH coupling $^2J_{CH}$ becomes more positive with increasing CCH bond angle and with decreasing electronegativity of the substituent on the coupling C. This property enables a distinction to be made inter alia between the substituents on the benzene ring

Table 2.6. Structural features (C-hybridisation, electronegativity, ring size) and typical one-bond CH coupling constants J_{CH} (Hz)[4-6]

Table 2.7. Structural features and *geminal* (two-bond) *CH* coupling constants, $^2J_{CH}$ (Hz)[4–6, 16]

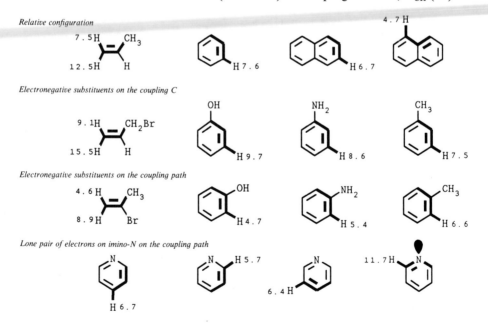

Table 2.8. Structural features and *vicinal* (three-bond) *CH* coupling constants, $^3J_{CH}$ (Hz)[4–6, 16]

or between heteroatoms in five-ring heteroaromatics (Table 2.7). From Table 2.7, those $^2J_{CH}$ couplings which may be especially clearly distinguished are:

β-C atoms in imines (e.g. C-3 in pyridine: 7 Hz);

α-C atoms in aldehydes (25 Hz);

quaternary C atoms of terminal ethynyl groups (40–50 Hz).

Vicinal CH couplings $^3J_{CH}$ depend not only on the configuration of the coupling C and *H* (Table 2.8; see Section 2.3.2), but also on the nature and position of substituents: an

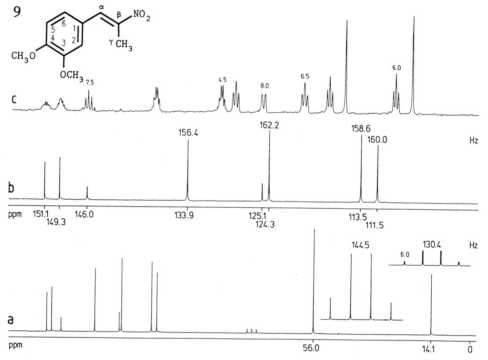

Fig. 2.9. ^{13}C NMR spectra of 3,4-dimethoxy-β-methyl-β-nitrostyrene (**9**) (CDCl₃, 25 °C, 20 MHz). (a, b) ^1H broadband decoupled, (a) with CH₃ quartets at 14.1 and 56.0 ppm; (c) coupled ('gated' decoupled). Assignments:

C	δ_C (ppm)		J_{CH} (Hz)	$^{3(2)}J_{CH}$ (Hz)		(coupling protons)
C-1	125.1	S		d	8.0	(5-H)
C-2	113.5	D	158.6	't'ᵃ	6.0	(6-H, α-H)
C-3	149.3	S		m		(3-H, 5-OCH₃)
C-4	151.1	S		m		(2-H, 6-H, 4-OCH₃)
C-5	111.5	D	160.0			
C-6	124.3	D	162.2	't'ᵃ	6.5	(2-H, α-H)
C-α	133.9	D	156.4	'sxt'ᵃ	4.5	(2-H, 6-H, β-CH₃)
C-β	146.0	S		'qui'ᵃ	7.5	(α-H, β-CH₃)
C-γ	14.1	Q	130.4	d	6.0	(α-H)
(OCH₃)₂	56.0	Q	144.5			

ᵃThe quotation marks indicate that the coupling contants are virtually the same for non-equivalent protons. C-β should, for example, split into a doublet ($^2J_{CH}$ to α-H) of quartets ($^2J_{CH}$ to β-CH₃); since both couplings have the same value (7.5 Hz), a pseudoquintet 'qui' is observed

electronegative substituent raises the $^3J_{CH}$ coupling constant on the coupled C and lowers it on the coupling path, e.g. in alkenes and benzene rings (Table 2.8). An imino-N on the coupling path (e.g. from C-2 to 6-H in pyridine, Table 2.8) is distinguished by a particularly large $^3J_{CH}$ coupling constant (12 Hz).

In the ^{13}C NMR spectra of benzene derivatives, apart from the $^1J_{CH}$, only the *meta* coupling ($^3J_{CH}$, but not $^2J_{CH}$) is resolved. A benzenoid CH, from whose perspective the *meta* positions are substituted, usually appears as a $^1J_{CH}$ doublet without additional splitting, e.g. in the case of 3,4-dimethoxy-β-methyl-β-nitrostyrene (**9**) (Fig. 2.9) the carbon atom C-5 generates a doublet at 111.5 ppm in contrast to C-2 at 113.5 ppm which additionally splits into a triplet. The use of *CH* coupling constants as criteria for assigning a resonance to a specific position is illustrated by this example.

Usually there is no splitting between two exchangeable XH protons (X = O, N, S) and C atoms through two or three bonds ($^2J_{CH}$ or $^3J_{CH}$), unless an intramolecular H bridge fixes the XH proton in the molecule. Thus the C atoms *ortho* to the hydroxy group show $^3J_{CH}$ coupling to the hydrogen bonding OH proton in salicylaldehyde (**10**), whose values reflect the relative configurations of the coupling partners. This method may be used, for example, to identify and assign the resonances in problem 15.

10   cis: 5.6 Hz   trans: 6.7 Hz

2.2.5 NH COUPLING CONSTANTS

Compared with ^1H and ^{13}C, the magnetic moment of ^{15}N is very small and has a negative value. The NH coupling constants are correspondingly smaller and their values are usually the reverse of comparable HH and CH couplings. Table 2.9 shows that the one-bond NH coupling, J_{NH}, is proportional to the s-character of the hybrid bonding orbital on N so a distinction can be made between amino- and imino-NH. Formamides can be identified by large $^2J_{NH}$ couplings between ^{15}N and the formyl proton. The $^2J_{NH}$ and $^3J_{NH}$ couplings of pyrrole and pyridine are especially distinctive and reflect the orientation of the non-bonding electron pair on nitrogen (pyrrole: perpendicular to the ring plane; pyridine: in the ring plane; Fig. 2.9), a fact which can be exploited in the identification of heterocyclic compounds (problems 24 and 25).

2.2.6 HH COSY (GEMINAL, VICINAL, w-RELATIONSHIPS OF PROTONS)

The *HH* COSY technique[12, 13, 17–19] in proton magnetic resonance is a quick alternative to spin decoupling[2, 3] in structure elucidation. 'COSY' is the acronym derived from COrrelation SpectroscopY. *HH* COSY correlates the ^1H shifts of the coupling protons of a molecule. The proton shifts are plotted on both frequency axes in the two-dimensional experiment. The result is a diagram with square symmetry (Fig. 2.10). The projection of the one-dimensional ^1H NMR spectrum appears on the diagonal (*diagonal signals*). In

Table 2.9. Structural features and typical NH coupling constants (Hz)[7]

addition there are *correlation* or *off-diagonal signals* (cross signals) where the protons are coupled with one another. Thus the *HH* COSY diagram indicates *HH* connectivities, that is, *geminal*, *vicinal* and *w*-relationships of the *H* atoms of a molecule and the associated structural units.

An *HH* COSY diagram can be shown in perspective as a stacked plot (Fig. 2.10a). Interpretation of this neat, three-dimensional representation, where the signal intensity gives the third dimension, can prove difficult because of distortions in the perspective. The contour plot can be interpreted more easily. This shows the signal intensity at various cross-sections (contour plots, Fig. 2.10b). However the choice of the plane of the cross-section affects the information provided by an *HH* COSY diagram; if the plane of the cross-section is too high then the cross signals which are weak are lost; if it is too low, then weaker artefacts may be mistaken for cross signals.

Every *HH* coupling interaction can be identified in the *HH* COSY contour plot by two diagonal signals and the two cross signals of the coupling partners, which form the four corners of a square. The coupling partner (cross signal) of a particular proton generates a signal on the vertical or horizontal line from the relevant ^1H signal. In Fig. 2.10b, for example, the protons at *7.90* and *7.16 ppm* are found as coupling partners on both the vertical and the horizontal lines from the proton *2-H* of quinoline (**11**) at *8.76 ppm*. Since *2-H* (*8.76 ppm*) and *3-H* (*7.16 ppm*) of the pyridine ring in **11** can be identified by the common coupling $^3J_{HH} = 5.5$ Hz (Table 2.5), the *HH* relationship which is likewise derived from the *HH* COSY diagram confirms the location of the pyridine protons in **11a**. Proton *4-H* of quinoline (*7.90 ppm*) shows an additional cross signal at *8.03 ppm* (Fig. 2.10). If it is known that this so-called zig-zag coupling is attributable to the benzene ring proton *8-H* (**11b**), then two further cross signals from *8.03 ppm* (at *7.55* and *7.35 ppm*) locate the remaining protons of quinoline (**11c**).

This example (Fig. 2.10) also shows the limitations of the *HH* COSY technique: first,

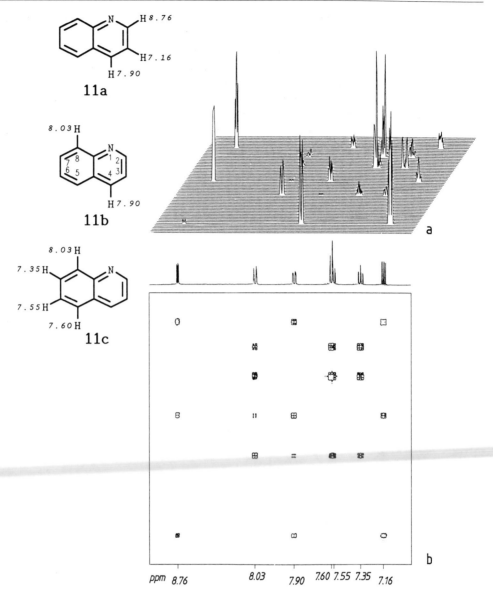

Fig. 2.10. *HH* COSY diagram of quinoline (**11**) [(CD$_3$)$_2$CO, 95% v/v, 25°C, 400 MHz, 256 scans]. (a) Stacked plot; (b) contour plot

evaluation, without taking known shifts and possible couplings into account, is not always conclusive because the cross-sectional area of the cross signals may not reveal which specific couplings are involved; second, overlapping signals (e.g. *7.55* and *7.60 ppm* in Fig. 2.10) are not separated by *HH* COSY if the relevant protons couple to one another. If there is sufficient resolution, however, the fine structure of the multiplets may be recognised by the shapes of the diagonal and cross signals, e.g. in Fig. 2.10, at *7.55 ppm* there is a triplet, therefore the resonance at *7.60 ppm* is a doublet (see the shape of the signal on the diagonal at *7.55–7.60 ppm* in Fig. 2.10).

In the case of *n*-fold splitting in one-dimensional ^1H NMR spectra the *HH* COSY diagram gives (depending on the resolution) up to n^2-fold splitting of the cross signals. If several small coupling constants contribute to a multiplet, the intensity of the cross signals in the *HH* COSY plot may be distributed into many multiplet signals so that even at a low-lying cross-section no cross signals appear in the contour diagram.

Hence the cross signals for the coupling partner of the bridgehead proton at *2.06 ppm* are missing in the *HH* COSY diagram of α-pinene (Fig. 2.11a) because the former is split

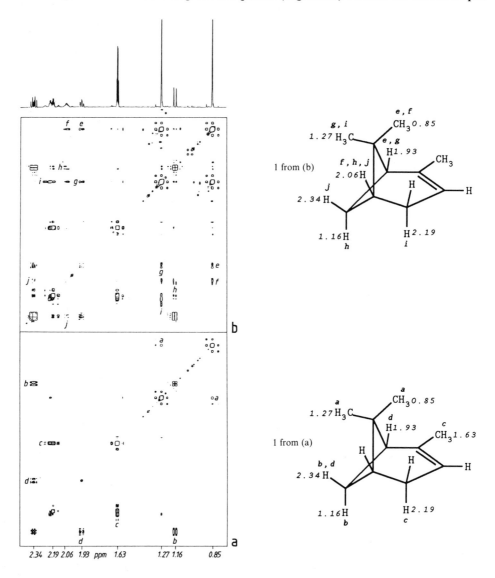

Fig. 2.11. *HH* COSY contour plot of α-pinene (**1**) [purity 98%, $(CD_3)_2CO$, 10% v/v, 25 °C, 400 MHz, 256 scans]. (a) Without time delay; (b) with time delay in the pulse sequence

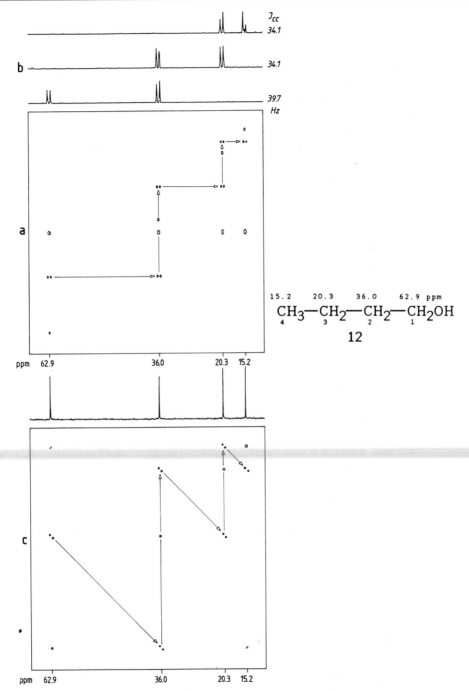

$$CH_3-CH_2-CH_2-CH_2OH$$

12

Fig. 2.12. Two-dimensional (2D-)INADEQUATE diagram of 1-butanol (**12**) [(CD$_3$)$_2$CO, 95% v/v, 25 °C, 50 MHz, 128 scans]. (a) Contour plot with the *AB* systems of bonded C atoms on the horizontal axis; (b) illustration of the three *AB* systems of the molecule from (a); (c) contour plot of the symmetrised INADEQUATE experiment showing the *AB* or *AX* systems of bonded C atoms in the *HH* cosy format (cross signals on the axes perpendicular to the diagonal)

into a multiplet with many weak individual signals as a result of several smaller couplings. In such cases, modifications of the *HH* COSY experiment such as "TOCSY" (from total correlation spectroscopy)[2] or "COSY with delay" are useful, the latter involving a delay which is adjusted to smaller couplings. Figure 2.11b shows that these variations of the *HH* COSY technique emphasise connectivities which are the result of smaller couplings. In the usual *HH* COSY diagram (Fig. 2.11a) are found only the *HH* connectivities a–d; a delay (Fig. 2.11b) gives the additional *HH* connectivities e–j.

Despite these limitations, structural fragments may almost always be derived by means of the *HH* COSY technique, so that with complementary information from other two-dimensional NMR experiments it is possible to deduce the complete structure. Thus the *HH* COSY technique is especially useful in finding a solution to problems 10, 28, 29, 45, 49 and 50.

2.2.7   CC INADEQUATE (CC BONDS)

Once all of the CC bonds in a molecule are known, then its carbon skeleton is established. One way to identify the CC bonds would be to measure ^{13}C–^{13}C coupling constants, since these are the same for C atoms which are bonded to one another: identical coupling constants are known to identify the coupling partners (see Section 2.2.3). Unfortunately, this method is complicated by two factors: first, ^{13}C–^{13}C couplings, especially those in the aliphatic region, are nearly all the same (35–40 Hz,[16] Fig. 2.12), provided that none of the coupling C atoms carries an electronegative substituent. Second, the occurrence of ^{13}C–^{13}C coupling requires the two nuclei to be directly bonded. However, given the low natural abundance of ^{13}C (1.1% or 10^{-2}), the probability of a ^{13}C—^{13}C bond is only 10^{-4}. Splitting as a result of ^{13}C–^{13}C coupling therefore appears only as a weak feature in the spectrum (0.5% intensity), usually in satellites which are concealed by noise at a distance of half the ^{13}C–^{13}C coupling constant on either side of the ^{13}C–^{12}C main signal (99% intensity).

The one-dimensional variations of the INADEQUATE experiment[12, 13, 17, 20] suppress the intense ^{13}C–^{13}C main signal, so that both *AX* and *AB* systems appear for all ^{13}C—^{13}C bonds in one spectrum. The two-dimensional variations[12, 13, 17, 21, 22] segregate these *AB* systems on the basis of their individual double quantum frequencies (the sum of the ^{13}C shifts of *A* and *B*) as a second dimension. Using the simple example of 1-butanol (**12**), Fig. 2.12a demonstrates the use of the two-dimensional INADEQUATE technique for the purpose of structure elucidation. For every C—C bond the contour diagram gives an *AB* system parallel to the abscissa with double quantum frequency as ordinate. By following the arrows in Fig. 2.12a, the carbon connectivities of butanol can be derived immediately. The individual *AB* systems may also be shown one-dimensionally (Fig. 2.12b); the ^{13}C–^{13}C coupling constants often provide useful additional information.

A variation on the INADEQUATE technique, referred to as symmetrised 2D INADEQUATE,[21,22] provides a representation in the *HH* COSY format with its quadratic symmetry of the diagonal and cross signals. Here the one-dimensional ^{1}H broadband decoupled ^{13}C NMR spectrum is projected on to the diagonal and the *AB* systems of all C—C bonds of the molecule are projected on to individual orthogonals (Fig. 2.12c). Every C—C bond then gives a square defined by diagonal signals and off-diagonal *AB* patterns, and it is possible to evaluate as described for *HH* COSY.

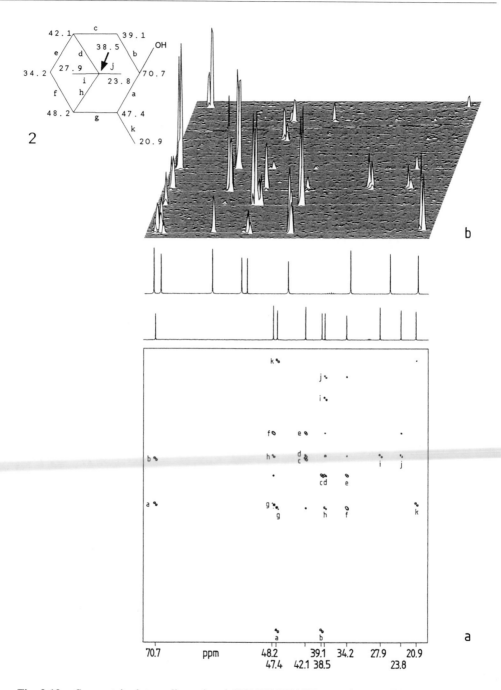

Fig. 2.13. Symmetrised two-dimensional INADEQUATE experiment with isopinocampheol (**2**) [(CD$_3$)$_2$CO, 250 mg in 0.3 ml, 25 °C, 50 MHz, 256 scans]. (a) Complete contour plot labelled a–k for the 11 CC bonds of the molecule to facilitate the assignments sketched in formula **2**; (b) stacked plot of the section between 20.9 and 48.2 ppm

A disadvantage is the naturally low sensitivity of the INADEQUATE technique. However, if one has enough substance (5–10 mg per C atom, samples from syntheses), then the sophisticated experiment is justified as the solutions to the problems 19, 20, 32 and 35 illustrate. Figure 2.13 is intended to demonstrate the potential of this technique for tracing out a carbon skeleton using the example of isopinocampheol (**2**). The evaluation of all CC–*AB* systems on the orthogonals leads to the eleven C–C bonds a–k. If all the C–C bonds which have been found are combined, then the result is the bicyclic system (a–h) and the three methyl substituents (i–k) of the molecule **2**. The point of attachment of O*H* group of the molecule (at *70.7 ppm*) is revealed by the DEPT technique in Fig. 2.5. Figure 2.13 also shows the *AB* effect on the ^{13}C signals of neighbouring C atoms with a small shift difference (bond g with *47.4* and *48.2 ppm*): the intense inner signals appear very clearly; the weak outer signals of the *AB* system of these two C atoms are barely recognisable except as dots. Additional cross signals without doublet structure, e.g. between *48.2* and *42.1 ppm*, are the result of longer range $^{2}J_{CC}$ and $^{3}J_{CC}$ couplings.

2.2.8   *CH* COSY (*CH* BONDS)

The *CH* COSY technique [12, 13, 17, 23] with carbon-13 detection and proton decoupling or a more sensitive method with ^{1}H detection and ^{13}C decoupling denoted as *inverse CH* COSY or HMQC (heteronuclear multiple quantum coherence) correlate ^{13}C shifts in one dimension with the ^{1}H shifts in the other via one-bond C*H* coupling J_{CH}.

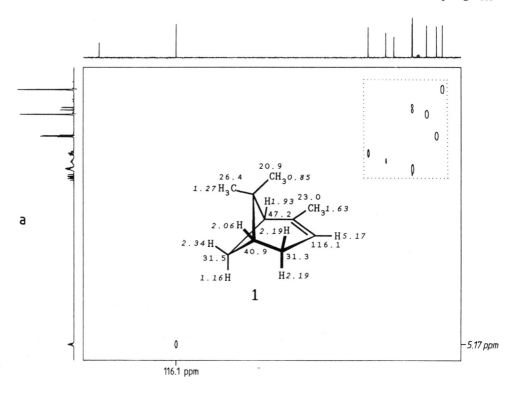

Fig. 2.14. (*caption opposite*)

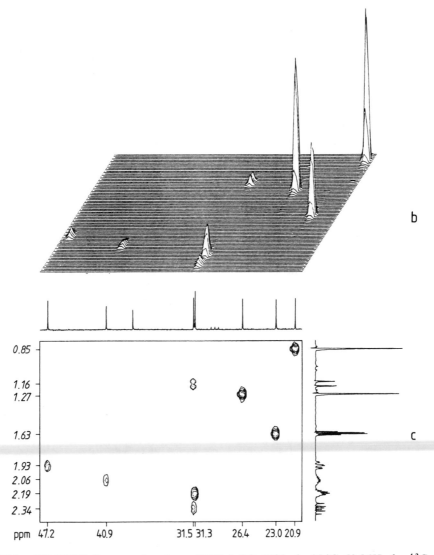

Fig. 2.14. *CH* COSY diagram of α-pinene [(CD$_3$)$_2$CO, 10% v/v, 25 °C, 50 MHz for ^{13}C, 200 MHz for ^{1}H, 256 scans]. (a) Complete contour plot; (b) stacked plot of the signals for the section outlined in (a) from 20.9 to 47.2 ppm (^{13}C) and *0.85 to 2.34 ppm* (^{1}H); (c) contour plot of section (b), showing one-dimensional ^{13}C and ^{1}H NMR spectra for this section aligned parallel to the abscissa and the ordinate

The pulse sequence which is used to record *CH* COSY also involves the ^{1}H– ^{13}C polarisation transfer which is the basis of the DEPT sequence and which increases the sensitivity by a factor of up to four. Consequently, a *CH* COSY experiment does not require any more sample than a ^{1}H broadband decoupled ^{13}C NMR spectrum. The result is a two-dimensional *CH* correlation, in which the ^{13}C shift is mapped on to the abscissa and the ^{1}H shift is mapped on to the ordinate (or vice versa). The ^{13}C and ^{1}H shifts of the ^{1}H and ^{13}C nuclei which are bonded to one another are read as coordinates of the cross signal as shown in the *CH* COSY stacked plot (Fig. 2.14b) and the associated contour plots of the α-pinene (Fig. 2.14a and c). To interpret them, one need only read off the coordinates of the correlation signals. In Fig. 2.14c, for example, the protons with shifts *1.16* (proton *A*) and *2.34 ppm* (proton *B* of an *AB* system) are bonded to the C atom at 31.5 ppm. The structural formula **1** shows all of the *CH* connectivities of α-pinene which can be read from Fig. 2.14.

The *CH* COSY technique is attractive because it is efficient and provides unequivocal results; it allows the shifts of two nuclei (^1H and ^{13}C) to be measured in a single experiment and within a feasible time scale. At the same time it determines all *CH* bonds (the *CH* connectivities) of the molecule, and hence provides an answer to the problem as to which ^1H nuclei are bonded to which ^{13}C nuclei. The fact that in the process the ^1H multiplets, which so frequently overlap in the ^1H dimension, are almost always separated in the second dimension (because of the larger frequency dispersion of the ^{13}C shifts) proves to be particularly advantageous especially in the case of larger molecules, a feature illustrated by the identification of several natural products (problems 40–50). The resolution of overlapping *AB* systems as in the case of ring CH_2 groups in steroids and in di- and triterpenes is especially helpful (problem 46). If there is sufficiently good resolution of the proton dimension in the spectrum, it may even be possible to recognise the fine structure of the ^1H multiplets from the shape of the correlation signals, a feature which is useful for solving problems 28, 42 and 46.

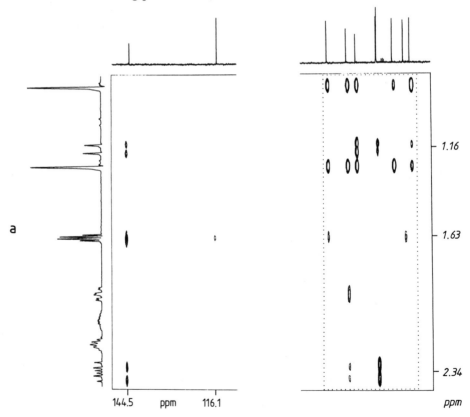

Fig. 2.15. (*caption opposite*)

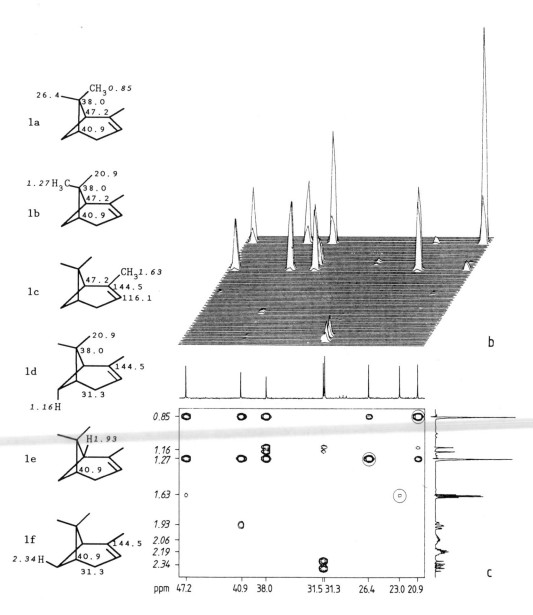

Fig. 2.15. *CH* COLOC investigation of α-pinene [$(CD_3)_2CO$, 10% v/v, 25 °C, 50 MHz for ^{13}C, 200 MHz for ^1H, 256 scans]. (a) Complete contour plot; (b) stacked plot of the section between 20.9 and 47.2 ppm (^{13}C) and *0.85 and 2.34 ppm* (^1H); (c) contour plot of (b). One-dimensional ^{13}C and ^1H NMR spectra for this section are shown aligned with the abscissa and ordinate of the contour plot. $^1J_{CH}$ correlation signals which are already known from the *CH* COSY study (Fig. 2.14) and have not been suppressed, are indicated by circles.

2.2.9   CH COLOC (*GEMINAL* AND *VICINAL* CH RELATIONSHIPS)

The CH COSY technique provides the $^1J_{CH}$ connectivities, and thereby applies only to those C atoms which are linked to H and not to quaternary C atoms. A modification of this technique, also applicable to quaternary C atoms, is one which is adjusted to the smaller $^2J_{CH}$ are $^3J_{CH}$ couplings (2–25 Hz, Tables 2.8 and 2.9).[23] Experiments that probe these couplings include CH COLOC [24] (correlation via long range couplings) with carbon-13 detection and HMBC (heteronuclear multiple bond coherence) with the more sensitive proton detection. These two-dimensional CH shift correlations indicate CH relationships through both two and three bonds ($^2J_{CH}$ and $^3J_{CH}$ connectivities) in addition to more or less suppressed $^1J_{CH}$ relationships which are in any case established from the CH COSY diagram. Format and analysis of the CH COLOC or HMBC plot correspond to those of a CH COSY or HMQC experiment, as is shown for α-pinene (1) in Fig. 2.15.

When trying to establish the CH relationships of a carbon atom (exemplified by the quaternary C at 38.0 ppm in Fig. 2.15), the chemical shift of protons at a distance of two or three bonds is found parallel to the ordinate (e.g. *0.85, 1.16* and *1.27 ppm* in Fig. 2.15). It is also possible to take the proton signals as the starting point and from the cross signals parallel to the abscissa to read off the shifts of the C atoms two or three bonds distant respectively. Thus, for example, one deduces that the methyl protons at *0.85* and *1.27 ppm* are two and three bonds apart from the C atoms at 38.0, 40.9 and 47.2 ppm as illustrated by the partial structures **1a** and **1b** in Fig. 2.15. CH correlation signals due to methyl protons prove to be especially reliable, as do *trans* CH relationships over three bonds, e.g. between *1.16* and 38.0 ppm in Fig. 2.15, in contrast to the missing *cis* relationship between *2.34* and 38.0 ppm.

2.3   Relative configuration and conformation

2.3.1   HH COUPLING CONSTANTS

Vicinal coupling constants $^3J_{HH}$ indicate very clearly the relative configuration of the coupling protons. Their contribution depends, according to the Karplus–Conroy equation:[2, 3]

$$^3J_{HH} = a\cos^2\Phi - 0.28 \quad \text{(up to } \Phi = 90°, a = 8.5; \text{ above } \Phi = 90°, a = 9.5) \qquad (1)$$

on the dihedral angle Φ, enclosed by the CH bonds as shown in Fig. 2.16, which sketches the Karplus–Conroy curves for dihedral angles from 0 to 180°. Experimental values are found between the two curves shown; electronegative substituents on the coupling path reduce the magnitude of $^3J_{HH}$.

$\Phi=60°$	$\Phi=180°$	$\Phi=-60°$
syn (gauche)	anti (trans)	−syn (gauche)
13a	**13b**	**13c**

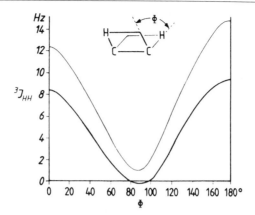

Fig. 2.16. Vicinal *HH* coupling constants $^3J_{HH}$ as a function of the dihedral angle Φ of the CH bonds concerned (Karplus–Conroy relationship). The lower, heavier, curve corresponds to the $\cos^2$ function given in the text. Experimental values lie between the two curves.

For the stable conformers **13a–c** of a substituted ethane the vicinal *HH* coupling constants $J_s \approx 3.5$ Hz for *syn*-protons and $J_a \approx 14$ Hz for *anti*-protons can be derived from Fig. 2.16. If there is free rotation around the C—C single bond, the coupling protons pass through the *syn* configuration twice and the *anti* configuration once. Therefore, from equation 2:

$$^3J_{\text{(average)}} = (2J_s + J_a)/3 = 21/3 = 7\,\text{Hz} \tag{2}$$

an average coupling constant of about *7 Hz* is obtained. This coupling constant characterises alkyl groups with unimpeded free rotation (cf. Figs 2.2 and 2.17).

Ethyl dibromodihydrocinnamate (**14**), for example, can form the three stable conformers **14a–c** by rotation around the CC single bond α to the phenyl ring.

Φ=60°	Φ=180°	Φ=–60°
syn (gauche)	*anti (trans)*	*–syn (gauche)*
14a	**14b**	**14c**

The ^1H NMR spectrum (Fig. 2.17) displays an *AB* system for the protons adjacent to this bond; the coupling constant $^3J_{AB} = 12$ Hz. From this can be deduced first that the dihedral angle Φ between the CH bonds is about 180°, second that conformer **14b** with minimised steric repulsion between the substituents predominates and third that there is restricted rotation around this CC bond. The relative configuration of the protons which is deduced from the coupling constant $^3J_{AB}$ confirms the conformation of this part of the structure of this molecule. On the other hand, the *HH* coupling constant of the ethyl

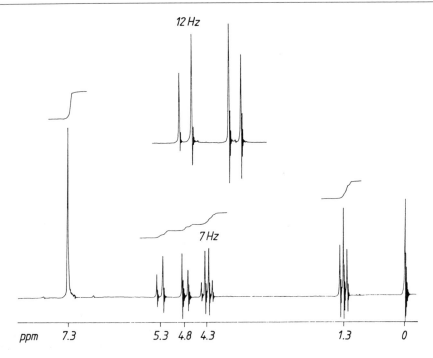

Fig. 2.17 ^1H NMR spectrum of ethyl dibromodihydrocinnamate (**14**) (CDCl$_3$, 25 °C, 90 MHz, CW recording)

group attached to oxygen (*7 Hz*, Fig. 2.17) reflects equal populations of all stable conformers around the CC bond of this ethyl group.

The $^3J_{HH}$ couplings shown in Table 2.10 verify the Karplus–Conroy equation 1 (Fig. 2.16) for rigid systems. Hence in cyclopropane the relationship $^3J_{HH(cis)} > {}^3J_{HH(trans)}$ holds, because *cis*-cyclopropane protons enclose a dihedral angle of about 0°, in contrast to an angle of *ca* 145° between *trans* protons, as shown by Dreiding models. Vicinal protons in cyclobutane, cyclopentane, norbornane and norbornene behave in an analogous way with larger *cis*, *endo-endo* and *exo-exo* couplings, respectively (Table 2.10).

Substituent effects (electronegativity, configuration) influence these coupling constants in four-, five- and seven-membered ring systems, sometimes reversing the *cis-trans* relationship,[2, 3] so that other NMR methods of structure elucidation, e.g. NOE difference spectra (see Section 2.3.5), are needed to provide conclusive results. However, the coupling constants of vicinal protons in cyclohexane and its heterocyclic analogues and also in alkenes (Table 2.10) are particularly informative.

Neighbouring *diaxial* protons of cyclohexane can be clearly identified by their large coupling constants ($^3J_{aa} \approx$ *11–13 Hz*, Table 2.10) which contrast with those of protons in *diequatorial* or *axial–equatorial* configurations ($^3J_{ee} \approx {}^3J_{ae} \approx$ *2–4 Hz*). Similar relationships hold for pyranosides as oxygen hetero analogues of cyclohexane, wherein the electronegative O atoms reduce the magnitude of the coupling constants ($^3J_{aa} \approx$ *9 Hz*, $^3J_{ae} \approx$ *4 Hz*, Table 2.10). These relationships are used for elucidation of the configuration of substituted cyclohexanes (problems 33–35), cyclohexenes (problems 10 and 32),

Table 2.10. $^3J_{HH}$ coupling constants (Hz) and relative configuration.[2, 3] The coupling path is shown in bold

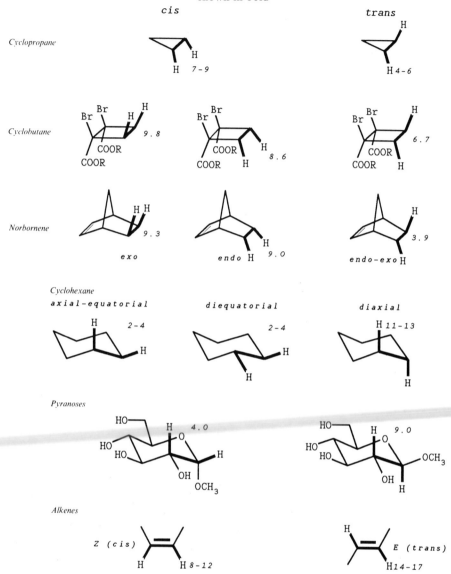

terpenes (problems 41, 42, 44 and 46), flavans (problem 8) and glycosides (problem 40, Table 2.10). In these cases also, the relative configuration of the protons which is deduced from the $^3J_{HH}$ coupling constant reveals the conformation of the six-membered rings. Thus the coupling constant *9 Hz* of the protons in positions 1 and 2 of the methyl-β-D-glucopyranoside **15** determines not only the *diaxial* configuration of the coupling protons but also the $^4C_1$ conformation of the pyranose ring. If the sterically more crowded $^1C_4$ conformation were present then a *diequatorial* coupling (*4 Hz*) of protons *1-H* and *2-H* would be observed. If the conformers were inverting (50:50 population of the $^4C_1$ and $^1C_4$ conformations), then the coupling constant would be the average (*6.5 Hz*).

The couplings of *vicinal* protons in 1,2-disubstituted alkenes lie in the range *6– 12 Hz* for *cis* protons (dihedral angle 0°) and *12–17 Hz* for *trans* protons (dihedral angle 180°), thus also following the Karplus-Conroy equation. Typical examples are the alkene proton *AB* systems of coumarin (**16a**) (*cis*) and *trans*-cinnamic acid (**16b**), and of the *cis-trans* isomers **17a** and **b** of ethyl isopentenyl ether, in addition to those in problems 3, 4, 9, 11 and 29.

2.3.2 *CH* COUPLING CONSTANTS

Geminal CH coupling constants $^2J_{CH}$ characterise the configuration of electronegative substituents in molecules with a defined geometry such as pyranose and alkenes.[16] If an electronegative substituent is attached *cis* with respect to the coupling proton, then the coupling constant $^2J_{CH}$ has a higher negative value; if it is located *trans* to the coupling proton, then $^2J_{CH}$ is positive and has a lower value; this is illustrated by β- and α-D-glucopyranose (**18a** and **b**) and by bromoethene (**19**).

Table 2.11. $^3J_{CH}$ coupling constants (Hz) and relative configuration.[16] The coupling path is shown in bold

Cyclohexane derivatives and pyranoses

cis

$\begin{matrix}H\\2.1\end{matrix}$

$\begin{matrix}CN & H \\ & 4.3\end{matrix}$

$\begin{matrix}H\\0-2\end{matrix}$

trans

$\begin{matrix}8.1\\H\end{matrix}$

$\begin{matrix}9.0\\H\end{matrix}$

CN

$\begin{matrix}5-6\\H\end{matrix}$

Alkenes

Carbon hybridisation

cis 7.5 H, CH₃

7.8 H

8.1 H

trans 12.6 H, H

11.9 H

14.7 H, CH₃

Electronegative substituents

cis 9.1 H, CH₂Br *on the coupling C*

4.6 H, CH₃ *on the coupling path*

trans 15.5 H, H

8.9 H, Br

Steric interactions

cis 10.1 H, =O

9.5 H, =O

CH₃, =O

trans 15.9 H, H

15.1 H, CH₃

11.0 H, CH₃

Vicinal CH coupling constants $^3J_{CH}$ resemble *vicinal HH* coupling constants in the way that they depend on the cosine^2 of the dihedral angle Φ between the CC bond to the coupled C atom and the *CH* bond to the coupled proton[16] (cf. Fig. 2.16), as illustrated by the Newman projections of the conformers **20a–c** of a propane fragment.

It follows from this that where there is free rotation about the CC single bond in alkyl groups then an averaged coupling constant $^3J_{CH}$ $(2J_{syn} + J_{anti})/3$ of between 4 and 5 Hz can be predicted, and that *vicinal CH* coupling constants $^3J_{CH}$ have values about two thirds of those of *vicinal* protons[16], $^3J_{HH}$.

Like $^3J_{HH}$ couplings, $^3J_{CH}$ couplings give conclusive information concerning the relative configuration of C and H as coupled nuclei in cyclohexane and pyranose rings and in alkenes (Table 2.11). Substituted cyclohexanes have $^3J_{CH} \approx 2$–4 Hz for *cis* and 8–9 Hz for *trans* configurations of the coupling partners; electronegative *OH* groups on the coupling path reduce the magnitude of $^3J_{CH}$ in pyranoses (Table 2.11). When deducing the configurations of multi-substituted alkenes, e.g. in solving problem 17, the $^3J_{CH}$ couplings of the alkenes in Table 2.11 are useful.

$^3J_{CH(trans)} > {}^3J_{CH(cis)}$ holds throughout. Electronegative substituents on the coupling carbon atom increase the J-value, whilst reducing it on the coupling path. Moreover, $^3J_{CH}$ reflects changes in the bonding state (C-hybridisation) and also steric hindrance (impeding coplanarity), as further examples in Table 2.11 show.

2.3.3   NH COUPLING CONSTANTS

The relationship between $^3J_{NH}$ and the dihedral angle of the coupling nuclei, of the type that applies to *vicinal* couplings of ^1H and ^{13}C, very rarely permits specific configurational assignments because the values ($^3J_{NH} < 5$ Hz) are too small.[7] In contrast, *geminal* couplings $^2J_{NH}$ distinguish the relative configurations of aldimines very clearly. Thus, *anti*-furan-2-aldoxime (**21a**) shows a considerably larger $^2J_{NH}$ coupling than does the *syn* isomer **21b**; evidently in imines the non-bonding electron pair *cis* to the *CH* bond of the coupled proton has the effect of producing a high negative contribution to the *geminal NH* coupling.

2.3.4 ^{13}C CHEMICAL SHIFTS

A C atom in an alkyl group is shielded by a substituent in the γ-position, that is, it experiences a smaller ^{13}C chemical shift or a negative substituent effect.[4–6] This originates from a sterically induced polarisation of the CH bond: the van der Waals radii of the substituent and of the hydrogen atom on the γ-C overlap; as a result, the σ-bonding electrons are moved from H towards the γ-C atom; the higher electron density on this C atom will cause shielding. As the Newman projections **22a–c** show, a distinction can be made between the stronger γ-*syn* and the weaker γ-*anti* effect. If there is free rotation, then the effects are averaged according to the usual expression, $(2\gamma_{syn} + \gamma_{anti})/3$, and one observes a negative γ-substituent effect of -2.5 to -3.5 ppm,[4–6] which is typical for alkyl groups.

In rigid molecules, strong γ-effects on the ^{13}C shift (up to 10 ppm) allow the different configurational isomers to be distinguished unequivocally, as *cis*- and *trans*-3- and -4-methylcyclohexanol (Table 2.12) illustrate perfectly: if the OH group is positioned *axial*, then its van der Waals repulsion of a *coaxial H* atom shields the attached C atom in the γ-position. 1,3-*Diaxial* relationships between substituents and H atoms in cyclohexane, norbornane and pyranosides shield the affected C atoms, generating smaller ^{13}C shifts than for isomers with *equatorial* substituents (Table 2.12).

The ^{13}C chemical shift thus reveals the relative configuration of substituents in molecules with a definite conformation, e.g. the *axial* position of the OH group in *trans*-3-methylcyclohexanol, *cis*-4-methylcyclohexanol, β-D-arabinofuranose and α-D-xylopyranose (Table 2.12). It turns out, in addition, that these compounds also take on the conformations shown in Table 2.12 (arabinopyranose, $^1C_4$; the others, $^4C_1$); if they occurred as the other conformers, then the OH groups on C-1 in these molecules would be *equatorial* with the result that larger shifts for C-1, C-3 and C-5 would be recorded. A ring inversion (50:50 population of both conformers) would result in an average ^{13}C shift.

Compared with ^1H chemical shifts, ^{13}C shifts are more sensitive to steric effects, as a comparison of the ^1H and the ^{13}C NMR spectra of *cis*- and *trans*-4-*tert*-butylcyclohexanol (**23**) in Fig. 2.18 shows. The polarisation through space of the γ-CH bond by the *axial OH* group in the *cis* isomer **23b** shields C-1 by -5.6 and C-3 by -4.8 ppm (γ-effect). In contrast the ^1H shifts reflect the considerably smaller anisotropic effect (see Section 2.5.1) of cyclohexane bonds: *equatorial* substituents (in this case H and OH) display larger shifts than *axial* substituents; the *equatorial 1-H* in *cis*-**23** (*3.92 ppm*) has a larger shift than the *axial 1-H* in *trans*-**23** (*3.40 ppm*); the difference is significantly smaller

Table 2.12. ^{13}C chemical shifts (ppm) and relative configurations of cycloalkanes, pyranoses and alkenes (application of γ-effects).[4-6] The shifts which are underlined reflect γ-effects on C atoms in the corresponding isomer pairs

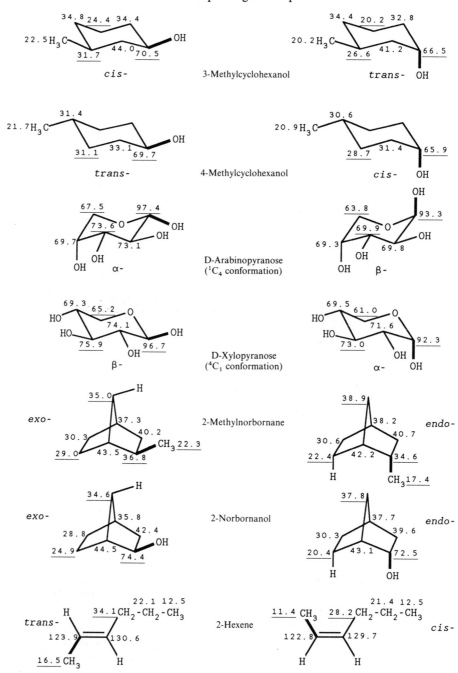

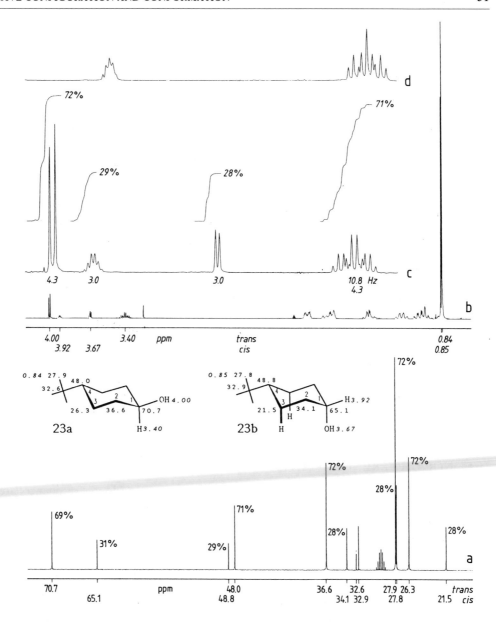

Fig. 2.18. NMR spectra of *trans-* and *cis*-4-*tert*-butylcyclohexanol (**23a** and **23b**) [$(CD_3)_2CO$, 25 °C, 400 MHz for ^1H, 100 MHz for ^{13}C]. (a) ^1H decoupled ^{13}C NMR spectrum (NOE suppressed, comparable signal intensities); (b) ^1H NMR spectrum; (c) section of (b) (*3–4 ppm*) with integrals; (d) partial spectrum (c) following D_2O exchange. The integrals (c) and the ^{13}C signal intensities (a) give the *trans*:*cis* isomer ratio 71:29. Proton *1-H* (*3.40 ppm*) in the *trans* isomer **23a** forms a triplet (*10.8 Hz*, two *anti* protons in 2, 2′-positions) of quartets (*4.3 Hz*, two *syn* protons in 2, 2′-positions and the O*H* proton as additional coupling partner); following D_2O exchange a triplet (*10.8 Hz*) of triplets (*4.3 Hz*) appears, because the coupling to O*H* is missing. In the *cis* isomer **23b** proton *1-H* forms a sextet (*3.0 Hz*, four *synclinal* protons in 2, 2′-positions and O*H*) which appears as a quintet following D_2O exchange because the coupling to O*H* is then lost.

($-0.52\,ppm$) than the γ-effect on the ^{13}C shifts (ca -5 ppm). Both spectra additionally demonstrate the value of NMR spectroscopy for quantitative analysis of mixtures by measuring integral levels or signal intensities, respectively. Finally, D_2O exchange eliminates the OH protons and their couplings from the ^1H NMR spectra (Fig. 2.18d).

The γ-effect on the ^{13}C shift also causes the difference between (E)- and (Z)-configurations of the alkyl groups in alkenes. Here the α-C atom shift responds most clearly to the double bond configurational change: these atoms in cis-alkyl groups occupy γ-positions with respect to each other; they enclose a dihedral angle of $0°$, and therefore are eclipsed, which leads to an especially strong van der Waals interaction and a correspondingly strong shielding of the ^{13}C nucleus. For this reason, the relationship $\delta_{trans} > \delta_{cis}$ holds for the α-C atoms of alkenes, as shown in Table 2.12 for (E)- and (Z)-2-hexene. The ^{13}C shifts of the doubly bonded carbon atoms behave similarly, although the effect is considerably smaller.

α,β-Unsaturated carbonyl compounds show smaller ^{13}C shifts than comparable saturated compounds,[4-6] provided that their carbonyl and CC double bonds are coplanar. If steric hindrance prevents coplanarity, conjugation is reduced and so larger ^{13}C shifts are observed. In α,β-unsaturated carbonyl compounds such as benzophenones and benzoic acid derivatives the twist angle θ between the carbonyl double bond and the remaining π-system can be read off and hence the conformation derived from the ^{13}C shift,[25] as several benzoic acid esters (24) illustrate.

R	$\delta_{C=O}$ (ppm)	θ °
H	166.9	0
CH_3	170.4	49
$CH(CH_3)_2$	171.3	57
$C(CH_3)_3$	173.1	90

2.3.5   NOE DIFFERENCE SPECTRA

Changes in signal intensities caused by spin decoupling (double resonance) are referred to as the nuclear Overhauser effect (NOE).[3, 26] In proton decoupling of ^{13}C NMR spectra, the NOE increases the intensity of the signals generated by the C atoms which are bonded to hydrogen by up to 200%; almost all techniques for measuring ^{13}C NMR spectra exploit this gain in sensitivity.[3, 6, 7] If in recording ^1H NMR spectra certain proton resonances are decoupled (homonuclear spin decoupling), then the changes in intensity due to the NOE are considerably smaller (much less than 50%).

For the assignment of configuration it is useful that, during excitation of a particular proton by spin decoupling, other protons in the vicinity may be affected although not necessarily coupled with this proton. As a result of molecular motion and the dipolar relaxation processes associated with it, the populations of energy levels of the protons change;[3, 26] their signal intensities change accordingly (NOE). For example, if the signal intensity of one proton increases during spin decoupling of another, then these protons must be positioned close to one another in the molecule, irrespective of the number of bonds which separate them.

NOE difference spectroscopy has proved to be a useful method for studying the spatial proximity of protons in a molecule.[27] In this experiment the ^1H NMR spectrum is recorded during the decoupling of a particular proton (measurement 1); an additional measurement with a decoupling frequency which lies far away (the 'off-resonance' experiment) but is otherwise subject to the same conditions, is then the basis for a comparison (reference measurement 2). The difference between the two measurements provides the NOE difference spectrum, in which only those signals are shown whose intensities are increased (positive signal) or decreased (negative signal) by NOE.

Figure 2.19 illustrates NOE difference spectroscopy with α-pinene (**1**): decoupling of the methyl protons at *1.27 ppm* (experiment c) gives a significant NOE on the proton at

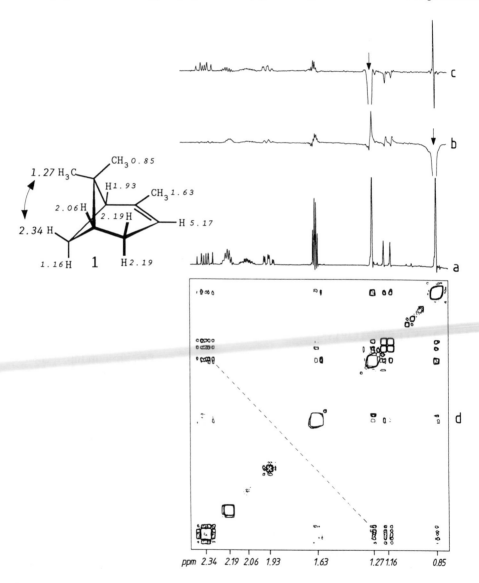

Fig. 2.19.   *HH* NOE difference spectra (b, c) and *HH* NOESY diagram (d) of α-pinene (**1**) with ^1H NMR spectrum (a) for comparison [(CD$_3$)$_2$CO, 10% v/v, 25 °C, 200 MHz, section from *0.85 to 2.34 ppm*)]. Vertical arrows in (b) and (c) indicate the decoupling frequencies; in the *HH* NOESY plot (d), cross-signals linked by a dotted line show the NOE detected in (c)

2.34 ppm; if for comparison, the methyl protons are decoupled at *0.85 ppm* (experiment b), then no NOE is observed at *2.34 ppm*. From this the proximity of the methyl-*H* atoms at *2.34 ppm* and the methyl group at *1.27 ppm* in α-pinene is detected. In addition, both experiments confirm the assignment of the methyl protons to the signals at *0.85* and *1.27 ppm*. A negative NOE, as on the protons at *1.16 ppm* in experiment c, is the result of coupling, e.g. in the case of the *geminal* relationship with the affected proton at *2.34 ppm*. Further applications of NOE difference spectroscopy are provided in problems 27, 30, 31, 42, 44, 46 and 48–50.

2.3.6 *HH* NOESY

The *HH* COSY sequence for ascertaining the *HH* connectivities also changes the populations of the energy levels, leading to NOEs. Thus, the *HH* COSY sequence has been expanded into the *HH* NOESY pulse sequence in order to give a two-dimensional measurement of changes in intensity.[28] The result of the measurements is shown in the *HH* NOESY plot with square symmetry (Fig. 2.19d) which is evaluated in the same way as *HH* COSY. Thus, Fig. 2.19d shows by the cross signals at *2.34* and *1.27 ppm* that the appropriate protons in α-pinene (**1**) are close to one another; the experiment also shows that the *HH* COSY cross signals (due to through-bond coupling) are not completely suppressed. Therefore, before evaluating an *HH* NOESY experiment, it is essential to know the *HH* connectivities from the *HH* COSY plot. A comparison of the two methods of NOE detection has shown that *HH* NOESY and its refinements (e.g. *HH* ROESY) are better suited to the investigation of the stereochemistry of biopolymers whereas for small- to medium-sized molecules (up to 50 C atoms) *HH* NOE difference spectroscopy is less time consuming, more selective and thus more conclusive.

2.4 Absolute configuration

2.4.1 DIASTEREOTOPISM

Where both *H* atoms of a methylene group cannot be brought into a chemically identical position by rotation or by any other movement of symmetry, they are said to be *diastereotopic*.[2, 3] The precise meaning of *diastereotopism* is best illustrated by means of an example, that of methylene protons H^A and H^B of glycerol (**25**). Where there is free rotation about the CC bonds, the terminal CH_2OH groups rotate through three stable conformations. They are best shown as Newman projections (**25a–c**) and the chemical environments of the CH_2O protons, H^A and H^B are examined with particular reference to *geminal* and *synclinal* neighbours.

H^A: OH, H^B; CH_2OH, H^C: δ_1
H^B: H^A, OH; OH, CH_2OH: δ_4

OH, H^B; H^C, OH: δ_2
H^A, OH; CH_2OH, H^C: δ_5

OH, H^B; OH, CH_2OH: δ_3
H^A, OH; H^C, OH: δ_6

It can be seen that the six possible near-neighbour relationships are all different. If rotation were frozen, then three different shifts would be measured for H^A and H^B in each of the conformations **a**, **b** and **c** (δ_1, δ_2 and δ_3 for H^A, δ_4, δ_5 and δ_6 for H^B). If there is free rotation at room temperature and if χ_a, χ_b and χ_c are the populations of conformations **a**, **b** and **c**, then according to the equations 3:

$$\delta_A = \chi_a\delta_1 + \chi_b\delta_2 + \chi_c\delta_3 \quad \text{and} \quad \delta_B = \chi_a\delta_4 + \chi_b\delta_5 + \chi_c\delta_6 \tag{3}$$

different average shifts $\delta_A \neq \delta_B$ are recorded which remain differentiated when all three conformations occur with equal population ($\chi_a = \chi_b = \chi_c = 1/3$). Chemical equivalence of such protons would be purely coincidental.

Figure 2.20 shows the diastereotopism of the methylene protons (CH^AH^BOH) of glycerol (**25**); it has a value of $\delta_B - \delta_A = 0.09\,ppm$. The spectrum displays an $(AB)_2C$ system for the symmetric constitution, $(CH^AH^BOH)_2H^COH$, of the molecule with *geminal* coupling $^2J_{AB} = 11.6\,Hz$ and the *vicinal* coupling constants $^3J_{AC} = 6.4$ and $^3J_{BC} = 4.5\,Hz$. The unequal ^3J couplings provide evidence against the unhindered free rotation about the CC bonds of glycerol and indicate instead that conformation **a** or **c** predominates with a smaller interaction of the substituents compared to **b**.

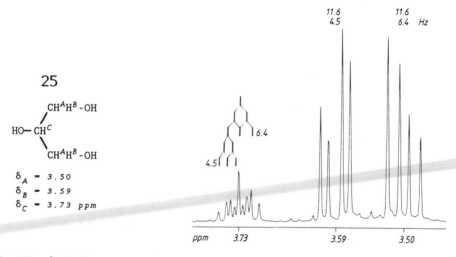

Fig. 2.20. ^1H NMR spectrum of glycerol (**25**) (D$_2$O, 10%, 25 °C, 400 MHz)

Diastereotopism indicates *prochirality*, as exemplified by glycerol (**25**) (Fig. 2.20). Other examples of this are diethylacetal, in which the OCH_2 protons are diastereotopic on account of the prochiral acetal-C atoms, thus forming AB systems of quartets (because of coupling with the methyl protons.)

The Newman projections **25a–c** draw attention to the fact that the central C atom, as seen from the terminal CH_2OH groups, appears asymmetric. It follows from this that diastereotopism is also a way of probing neighbouring asymmetric C atoms. Thus the methyl groups of the isopropyl residues in D- or L-valine (**26**) are diastereotopic and so show different ^1H and ^{13}C shifts, although these cannot be individually assigned to the two groups. In chiral alcohols of the type **27** the diastereotopism of the isopropyl-C nucleus increases with the size of the alkyl residues (methyl < isopropyl < *tert*-butyl).[29]

If a molecule contains several asymmetric C atoms, then the diastereomers show diastereotopic shifts. Clionasterol (**28a**) and sitosterol (**28b**) for example, are two steroids that differ only in the absolute configuration at one carbon atom, C-24.[30] Differing shifts of ^{13}C nuclei close to this asymmetric C atom in **28a** and **b** identify the two diastereomers including the absolute configuration of C-24 in both. The absolute configurations of carboxylic acids in pyrrolizidine ester alkaloids are also reflected in diastereotopic ^1H and ^{13}C shifts,[31] which is useful in solving problem 49.

28a Clionasterol (*24S*)   **28b** Sitosterol (*24R*)

2.4.2   CHIRAL SHIFT REAGENTS (*ee* DETERMINATION)

The presence of asymmetric C atoms in a molecule may, of course, be indicated by diastereotopic shifts and absolute configurations may, as already shown, be determined empirically by comparison of diastereotopic shifts.[30, 31] However, enantiomers are not differentiated in the NMR spectrum. The spectrum gives no indication as to whether a chiral compound exists in a racemic form or as a pure enantiomer.

Nevertheless, it is possible to convert a racemic sample with chiral reagents into diastereomers or simply to dissolve it in an enantiomerically pure solvent *R* or *S*; following this process, solvation diastereomers arise from the racemate (*RP* + *SP*) of the sample P, e.g. *R:RP* and *R:SP*, in which the enantiomers are recognisable because of their different shifts. Compounds with groups which influence the chemical shift because of their anisotropy effect (see Sections 2.5.1 and 2.5.2) are suitable for use as chiral solvents, e.g. 1-phenylethylamine and 2,2,2-trifluoro-1-phenylethanol.[32]

A reliable method of checking the enantiomeric purity by means of NMR uses europium(III) or praseodymium(III) chelates of type **29** as chiral shift reagents.[33] With a racemic sample, these form diastereomeric europium(III) or praseodymium(III) chelates, in which the shifts of the two enantiomers are different. Different signals for *R* and *S* will be observed only for those nuclei in immediate proximity to a group capable of coordination (*OH*, *NH$_2$*, C=O). The separation of the signals increases with increasing concentration of the shift reagent; unfortunately, line broadening of signals due to the paramagnetic ion increases likewise with an increase in concentration, which limits the

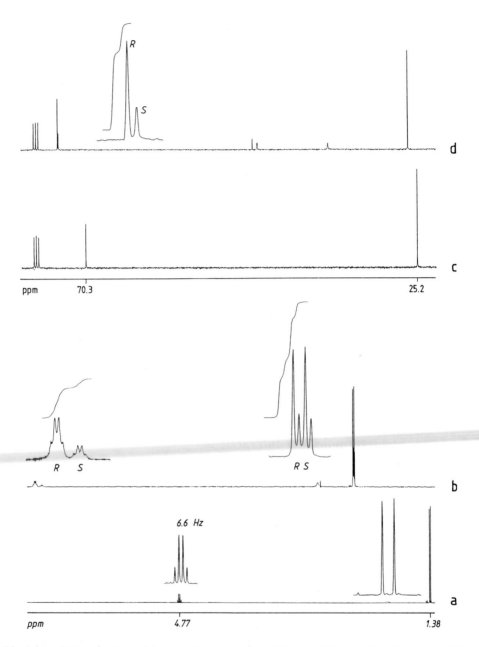

Fig. 2.21. Determination of the enantiomeric excess of 1-phenylethanol (**30**) (0.1 mmol in 0.3 ml $CDCl_3$, 25 °C) by addition of the chiral praseodymium chelate **29b** (0.1 mmol). (a, b) ^1H NMR spectra (400 MHz), (a) without the shift reagent and (b) with the shift reagent **29b**. In the ^{13}C NMR spectrum (d) only the α-C atoms of enantiomers **30R** and **30S** are resolved. The ^1H and ^{13}C signals of the phenyl residues are not shifted; these are not shown for reasons of space. The evaluation of the integrals gives 73% R and 27% S, i.e. an enantiomeric excess (*ee*) of 46%

amount of shift reagent which may be used. Figure 2.21 shows the determination of an enantiomeric excess (*ee*) following the equation

$$ee = \frac{R - S}{R + S} \times 100 \ (\%) \tag{4}$$

for 1-phenylethanol (**30**) by ^1H and ^{13}C NMR, using tris[3-(heptafluoropropylhydroxy-methylene)-D-camphorato]praseodymium(III) (**29b**) as a chiral shift reagent.

2.5   Intra- and intermolecular interactions

2.5.1   ANISOTROPIC EFFECTS

The chemical shift of a nucleus depends in part on its spatial position in relation to a bond or a bonding system. The knowledge of such anisotropic effects is useful in structure elucidation. An example of the anisotropic effect would be the fact that *axial* nuclei in cyclohexane almost always show smaller ^1H shifts than equatorial nuclei on the same C atom (illustrated in the solutions to problems 33–35, 41, 42, 44 and 46). The γ-effect also contributes to the corresponding behaviour of ^{13}C nuclei (see Section 2.3.4).

Multiple bonds are revealed clearly by anisotropic effects. Textbook examples include alkynes, shielded along the C≡C triple bond, and alkenes and carbonyl compounds, where the nuclei are deshielded in the plane of the C=C and C=O double bonds, respectively. One criterion for distinguishing methyl groups attached to the double bond of pulegone (**31**), for example, is the carbonyl anisotropic effect.

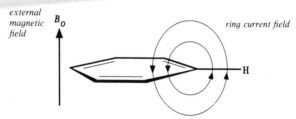

Alkynes   Alkenes   Aldehydes   **31**

2.5.2 RING CURRENT OF AROMATIC COMPOUNDS

Benzene shows a considerably larger ^1H shift (*7.28 ppm*) than alkenes (cyclohexene, *5.59 ppm*) or cyclically conjugated polyenes such as cyclooctatetraene (*5.69 ppm*). This is generally explained by the deshielding of the benzene protons by a ring current of π-electrons.[2, 3] This is induced when an aromatic compound is subjected to a magnetic field. The ring current itself produces its own magnetic field, opposing the external field within and above the ring, but aligned with it outside.[2, 3] As a result, nuclei inside or above an aromatic ring display a smaller shift whereas nuclei outside the ring on a level with it show a larger shift. The ring current has a stronger effect on the protons attached to or in the ring than on the ring C atoms themselves, so that particularly ^1H shifts prove a useful means of investigating ring currents and as aromaticity criteria for investigating annulenes.

Ring current model for benzene

1,4-Decamethylenebenzene (**32**) illustrates the ring current of benzene by a shielding of the methylene protons which lie above the aromatic ring plane in the molecule. A clear representation of the ring current effect is given by [18]annulene (**33**) at low temperature and the vinylogous porphyrin **35** with a diaza[26]annulene perimeter:[34] the inner protons are strongly shielded (*−2.88* and *−11.64 ppm*, respectively); the outer protons are strongly deshielded (*9.25* and *13.67 ppm*, respectively). The typical shift of the inner N*H* protons (*−2* to *−3 ppm*) indicates that porphyrin **34** occurs as diaza[18]annulene tautomers. Problem 34 draws attention to this point and offers the shielding ring current effect above the porphyrin ring plane for an analysis of the conformation.

2.5.3   INTRA- AND INTERMOLECULAR HYDROGEN BONDING

Hydrogen bonding can be recognised in ^1H NMR spectra by the large shifts associated with it; these large shifts are caused by the electronegativity of the heteroatoms bridged by the hydrogen atom. The OH protons of enol forms of 1,3-diketones are an extreme example. They form an *intramolecular H* bond and appear between *12.5 ppm* (hexafluoroacetylacetone enol, Fig. 2.22) and *15.5 ppm* (acetylacetone enol).

Intermolecular H bonding can be recognised in the ^1H NMR spectrum by the fact that the shifts due to the protons concerned depend very strongly on the concentration, as the simple case of methanol (**36**) demonstrates (Fig. 2.22a); solvation with tetrachloromethane as a solvent breaks down the *H* bridging increasingly with dilution of the solution; the OH shift decreases in proportion to this. In contrast, the shift of the ^1H signal of an *intramolecular* bridging proton remains almost unaffected if the solution is diluted as illustrated in the example of hexafluoroacetylacetone (**37**), which is 100% enolised (Fig. 2.22b).

Intermolecular H bonding involves an exchange of hydrogen between two heteroatoms in two different molecules. The H atom does not remain in the same molecule but

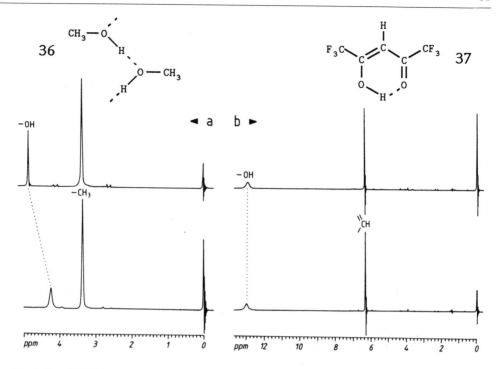

Fig. 2.22. ^1H NMR spectra of methanol (**36**) (a) and hexafluoroacetylacetone (**37**) (b), both in the pure state (above) and diluted in tetrachloromethane solution (5%, below) (25 °C, 90 MHz, *CW* recording)

is exchanged. If its exchange frequency is greater than as given by the Heisenberg uncertainty principle (equation 5),

$$v_{\text{exchange}} \geq \pi J_{AX}/\sqrt{2} \approx 2.22 J_{AX} \tag{5}$$

then its coupling J_{AX} to a *vicinal* proton H^A is not resolved. Hence CH_n protons do not generally show splitting by *vicinal* SH, OH or NH protons at room temperature. The same holds for $^3J_{CH}$ couplings with such protons. If the hydrogen bonding is *intramolecular*, then coupling is resolved, as the example of salicylaldehyde (**10**) has already shown (see Section 2.2.4; for an application, see problem 15).

2.5.4 PROTONATION EFFECTS

If a sample contains groups that can take up or lose a proton, H^+ (NH_2, $COOH$), then one must expect the pH and the concentration to affect the chemical shift when the experiment is carried out in an acidic or alkaline medium to facilitate dissolution. The pH may affect the chemical shift of more distant, nonpolar groups, as shown by the amino acid alanine (**38**) in neutral (betaine form **38a**) or alkaline solution (anion **38b**). The dependence of shift on pH follows the path of the titration curves; it is possible to read off the pK value of the equilibrium from the point of inflection.[2, 6]

^{13}C shifts respond to pH changes with even greater sensitivity; this is demonstrated by the values of pyridine (**39b**) and its cation (**39a**).

^{13}C shifts respond to pH changes with even greater sensitivity; this is demonstrated by the values of pyridine (**39b**) and its cation (**39a**).

The effect of pH is rarely of use for pK measurement; it is more often of use in identifying the site of protonation/deprotonation when several basic or acidic sites are present. Knowing the incremental substitutent effects $Z^{5, 6}$ of amino and ammonium groups on benzene ring shifts in aniline and in the anilinium ion (**40**), one can decide which of the N atoms is protonated in procaine hydrochloride (problem 21).

^{13}C chemical shifts relative to benzene (128.5 ppm) as reference

2.6   Molecular dynamics (fluxionality)

2.6.1   TEMPERATURE-DEPENDENT NMR SPECTRA

Figure 2.23 shows the ^1H NMR spectrum of N,N-dimethylacetamide (**41**) and its dependence on temperature. At 55 °C and below two resonances appear for the two N-methyl groups. Above 55 °C the signals become increasingly broad until they merge to form one broad signal at 80 °C. This temperature is referred to as the *coalescence temperature*, T_c. Above T_c the signal, which now belongs to both N-methyl groups, becomes increasingly sharp.

The temperature-dependent position and profile of the N-methyl signal result from amide canonical formulae of **41** shown in Fig. 2.23: the CN bond is a partial double bond; this hinders rotation of the N,N-dimethylamino group. One methyl group is now *cis* ($\delta_B = 3.0\,ppm$) and the other is *trans* ($\delta_A = 2.9\,ppm$) to the carboxamide oxygen. At low temperatures (55 °C), the N-methyl protons slowly exchange positions in the molecule (slow rotation, *slow exchange*). If energy is increased by heating (to above 90 °C), then the N,N-dimethylamino group rotates so that the N-methyl protons exchange their

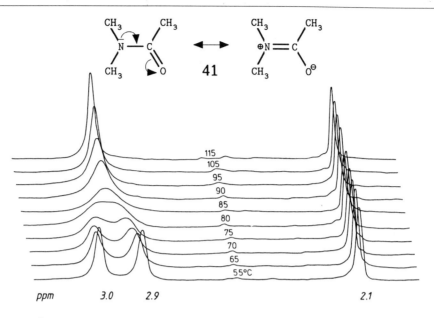

Fig. 2.23. ^1H NMR spectra of N, N-dimethylacetamide (**41**) at the temperatures indicated [$(CD_3)_2SO$, 75% v/v, 80 MHz]

position with a high frequency (free rotation, *rapid exchange*), and one single, sharp N-methyl signal of double intensity appears with the average shift ($\delta_B + \delta_A$)/2 = *2.95 ppm*.

The dimethylamino group rotation follows a first-order rate law; the exchanging methyl protons show no coupling and their singlet signals are of the same intensity. Under these conditions, equation 6[2, 35–37] affords the rate constant k_r at the coalescence point T_c:

$$k_r = \pi(\nu_B - \nu_A)/\sqrt{2} = \pi\,\Delta\nu/\sqrt{2} \approx 2.22\Delta\nu \qquad (6)$$

where $\Delta\nu$ is the full width at half-maximum of the signal at the coalesence point T_c; it corresponds to the difference in chemical shift ($\nu_B - \nu_A$) observed during slow exchange. In the case of dimethylacetamide (**41**) the difference in the chemical shift is *0.1 ppm* (Fig. 2.23), i.e. *8 Hz* (at 80 MHz). From equation 6 it can then be calculated that the N-methyl groups at the coalescence point (80 °C or 353 K) rotate with an exchange frequency of $k_r = 2.22 \times 8 \approx 17.8$ Hz.

According to the Eyring equation 7, the exchange frequency k_r decreases exponentially with the free molar activation enthalpy ΔG:[35–37]

$$k_r = \frac{kT_c}{h}\,e^{-\Delta G/RT_c} \qquad (7)$$

where R is the gas constant, k is the Boltzmann constant and h is Planck's constant. Equations 6 and 7 illustrate the value of temperature-dependent NMR for the investigation of molecular dynamics: following substitution of the fundamental constants,

Table 2.13. Selected applications of dynamic proton resonance[35-37]

			T_c (°K)	ΔG_{Tc} (kJ/mol)
Rotation hindered by bulky substituents (t-butyl groups)			147	30
Inversion at amino-nitrogen (aziridine)			380	80
Ring inversion (cyclohexane)			193	25
Valence tautomerism (Cope systems, fluctionality)			298	3

they give equation 8 for the free molar activation enthalpy ΔG for first-order exchange processes:

$$\Delta G = 19.1 T_c [10.32 + \log (T_c/k_r)] \times 10^{-3} \text{ kJ/mol} \qquad (8)$$

Hence the activation energy barrier to dimethylamino group rotation in dimethylaceta-mide (**41**) is calculated from equation 8 with $k_r = 17.8 \text{ s}^{-1}$ at the coalescence point 353 K (Fig. 2.23):

$$\Delta G_{353} = 78.5 \text{ kJ/mol or } 18.7 \text{ kcal/mol}$$

Temperature-dependent (dynamic) NMR studies are suited to the study of processes with rate constants between 10^{-1} and 10^3 s^{-1}.[3] Some applications are shown in Table 2.13 and in problems 11 and 12.

2.6.2 ^{13}C SPIN-LATTICE RELAXATION TIMES

The spin-lattice relaxation time T_1 is the time constant with which an assembly of a particular nuclear spin in a sample becomes magnetised parallel to the magnetic field as it is introduced into it. The sample magnetisation M_o is regenerated after every excitation with this time constant. For organic molecules the T_1 values of even differently bonded protons in solution are of the same order of magnitude (0.1–10 s). ^{13}C nuclei behave in a way which shows greater differentiation between nuclei and generally take more time: in molecules of varying size and in different chemical environments the spin-lattice relaxation times lie between a few milliseconds (macromolecules) and several minutes (quaternary C atoms in small molecules). Since during ^{1}H broadband decoupling only one T_1 value is recorded for each C atom (rather than n T_1 values as for all n components of a complex ^{1}H multiplet), the ^{13}C spin-lattice relaxation times are useful parameters for probing molecular mobility in solution.

The technique for measurement which is most easily interpreted is the inversion–recovery method,[3–6] in which the distribution of the nuclear spins among the energy levels is inverted by means of a suitable 180° radiofrequency pulse. A negative signal is observed at first, which becomes increasingly positive with time (and hence also with increasing spin–lattice relaxation) and which finally approaches the equilibrium intensity asymptotically. Figures 2.24 and 2.25 show asymptotical increases in the signal amplitude due to ^{13}C spin–lattice relaxation up to the equilibrium value using two instructive examples. A simple analysis makes use of the 'zero intensity interval', τ_o, without consideration of standard deviations: after this time interval τ_o, the spin–lattice relaxation is precisely far enough advanced for the signal amplitude to pass through zero. Equation 9 then gives T_1 for each individual C atom.

$$T_1 = \tau_o/\ln 2 \approx 1.45\,\tau_o \tag{9}$$

Thus, in the T_1 series of measurements of 2-octanol (**42**) (Fig. 2.24) for the methyl group at the hydrophobic end of the molecule, the signal intensity passes through zero at $\tau_o = 3.8$ s. From this, using equation 9, a spin–lattice relaxation time of $T_1 = 5.5$ s can be

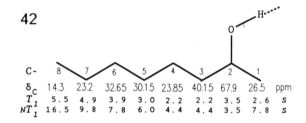

C-	8	7	6	5	4	3	2	1	
δ_C	14.3	23.2	32.65	30.15	23.85	40.15	67.9	26.5	ppm
T_1	5.5	4.9	3.9	3.0	2.2	2.2	3.5	2.6	s
NT_1	16.5	9.8	7.8	6.0	4.4	4.4	3.5	7.8	s

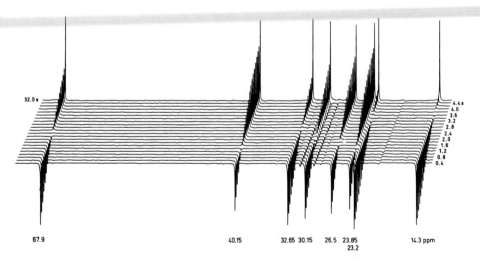

Fig. 2.24. Sequence of measurements to determine the ^{13}C spin–lattice relaxation times of 2-octanol (**42**) [(CD$_3$)$_2$CO, 75% v/v, 25 °C, 20 MHz, inversion–recovery sequence, stacked plot]. The times at which the signals pass through zero, τ_0, have been used to calculate (by equation 9) the T_1 values shown above for the ^{13}C nuclei of 2-octanol

calculated. A complete relaxation of this methyl C atom requires about five times longer (more than 30 s) than is shown in the last experiment of the series (Fig. 2.24); T_1 itself is the time constant for an exponential increase, in other words, after T_1 the difference between the observed signal intensity and its final value is still $1/e$ of the final amplitude.

The main contribution to the spin–lattice relaxation of ^{13}C nuclei which are connected to hydrogen is provided by the dipole–dipole interaction (DD mechanism, *dipolar relaxation*). For such ^{13}C nuclei a nuclear Overhauser enhancement of almost 2 will be observed during ^1H broadband decoupling according to equation 10:

$$\eta_c = \gamma_H/2\gamma_c = 1.988 \tag{10}$$

where γ_H and γ_c are the gyromagnetic constants of ^1H and ^{13}C.

If smaller NOE enhancements are recorded for certain ^{13}C nuclei, then other mechanisms (e.g. spin–rotation) contribute to their spin–lattice relaxation.[5,6]

Dipolar relaxation of ^{13}C nuclei originates from the protons (larger magnetic moment) in the same or in neighbouring molecules, which move with molecular motion (translation, vibration, rotation). This motion generates fluctuating local magnetic fields which affect the observed nucleus. If the frequency of a local magnetic field matches the Larmor frequency of the ^{13}C nucleus being observed (resonance condition), then this nucleus can undergo transition from the excited state to the ground state (relaxation) or the reverse (excitation). From this, it follows that the spin–lattice relaxation is linked to the mobility of the molecule or molecular fragment. If the average time taken between two reorientations of the molecule or fragment is defined as the correlation time τ_c, and if $n\,H$ atoms are connected to the observed C, then the dipolar relaxation time $T_{1(DD)}$ is given by the correlation function 11:

$$T_{1(DD)}^{-1} = \text{constant} \times n\tau_c \tag{11}$$

Accordingly, the relaxation time of a C atom will increase the fewer hydrogen atoms it bonds to and the faster the motion of the molecule or molecular fragment in which it is located. From this, it can be deduced that the spin–lattice relaxation time of ^{13}C nuclei provides information concerning four molecular characteristics:

Molecular size: smaller molecules move more quickly than larger ones; as a result, C atoms in small molecules relax more slowly than those in large molecules. The C atoms in the more mobile cyclohexanes ($T_1 = 19$–20 s) take longer than those in the more sluggish cyclodecane ($T_1 = 4$–5 s).[5, 6]

The number of bonded H atoms: if all parts within a molecule move at the same rate (the same τ_c for all C atoms), the relaxation times T_1 decrease from CH via CH_2 to CH_3 in the ratio given by equation 12:

$$T_1(CH):T_1(CH_2):T_1(CH_3) = 6:3:2 \tag{12}$$

Since methyl groups also rotate freely in otherwise rigid molecules, they follow the ratio shown in equation 12 only in the case of considerable steric hindrance.[6] In contrast, the T_1 values of ^{13}C nuclei of CH and CH_2 groups follow the ratio 2:1 even in large, rigid molecules. Typical examples are steroids such as cholesteryl chloride (**43**), in which the CH_2 groups of the ring relax at approximately double the rate (0.2–0.3 s) of CH carbon atoms (0.5 s). Contrary to the prediction made by equation 12, freely rotating methyl groups require considerably longer (1.5 s) for spin–lattice relaxation.

43 T_1 (s)

Segmental mobility: if one examines the T_1 series of 2-octanol (**42**) (Fig. 2.24) calculated according to equation 9, it becomes apparent that the mobility parameters nT_1 increase steadily from C-2 to C-8. As a result of hydrogen bonding, the molecule close to the OH groups is almost rigid (nT_1 between 3.5 and 4.4 s). With increasing distance from the anchoring effect of the OH group the mobility increases; the spin–lattice relaxation time becomes correspondingly longer. The nT_1 values of the two methyl groups also reflect the proximity to (7.8 s) and distance from (16.5 s) the hydrogen bond as a 'braking' device.

Anisotropy of molecular movement: monosubstituted benzene rings, e.g. phenyl benzoate (**44**), show a very typical characteristic: in the *para* position to the substituents the CH nuclei relax considerably more rapidly than in the *ortho* and *meta* positions. The reason for this is the anisotropy of the molecular motion: the benzene rings rotate more easily around an axis which passes through the substituents and the *para* position, because this requires them to push aside the least number of neighbouring molecules. This rotation, which affects only the *o*- and *m*-CH units, is too rapid for an effective spin–lattice relaxation of the *o*- and *m*-C atoms. More efficient with respect to relaxation are the frequencies of molecular rotations perpendicular to the preferred axis, and these affect the *p*-CH bond. If the phenyl rotation is impeded by bulky substituents, e.g. in 2,2′, 6,6′-tetramethylbiphenyl (**45**), then the T_1 values of the CH atoms can be even less easily distinguished in the *meta* and *para* positions (3.0 and 2.7 s, respectively).

44

45

46 T_1 (s)

Figure 2.25 shows the anisotropy of the rotation of the pyridine ring in nicotine (**46**). The main axis passes through C-3 and C-6; C-6 relaxes correspondingly more rapidly (3.5 s) than the three other CH atoms (5.5 s) of the pyridine ring in nicotine, as can be seen from the times at which the appropriate signals pass through zero.

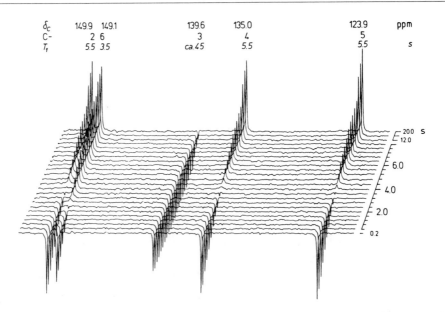

δ_c	149.9	149.1		139.6	135.0		123.9	ppm
C-	2	6		3	4		5	
T_1	5.5	3.5		ca.4.5	5.5		5.5	s

Fig. 2.25. Sequence of measurements to determine the spin–lattice relaxation times of the ^{13}C nuclei of the pyridine ring in L-nicotine (**46**) [$(CD_3)_2CO$, 75% v/v, 25 °C, inversion–recovery sequence, 20 MHz]. The times at which signals pass through zero have been used to calculate (by equation 9) the T_1 values for the pyridine C atoms in L-nicotine

2.7 Summary

Table 2.14 summarizes the steps by which molecular structures can be determined using the NMR methods discussed thus far to determine the structure, relative configuration and conformation of a specific compound.

In the case of completely unknown compounds, the molecular formula is a useful source of additional information; it can be determined using small amounts of substance (a few micrograms) by high-resolution mass spectrometric determination of the accurate molecular mass. It provides information concerning the double-bond equivalents (the 'degree of unsaturation'—the number of multiple bonds and rings).

For the commonest heteroatoms in organic molecules (nitrogen, oxygen, sulphur, halogen), the number of double-bond equivalents can be derived from the molecular formula by assuming that oxygen and sulphur require no replacement atom, halogen may be replaced by hydrogen and nitrogen may be replaced by CH. The resulting empirical formula C_nH_x is then compared with the empirical formula of an alkane with n C atoms, C_nH_{2n+2}; the number of double-bond equivalents is equal to half the hydrogen deficit, $(2n + 2 - x)/2$. From C_8H_9NO (problem 4), for example, the empirical formula C_9H_{10} is derived and compared with the alkane formula C_9H_{20}; a hydrogen deficit of ten and thus of five double-bond equivalents is deduced. If the NMR spectra have too few signals in the shift range appropriate for multiple bonds, then the double-bond equivalents indicate rings (see, for example, α-pinene, Fig. 2.4).

Table 2.14. Suggested tactics for solving structures using NMR

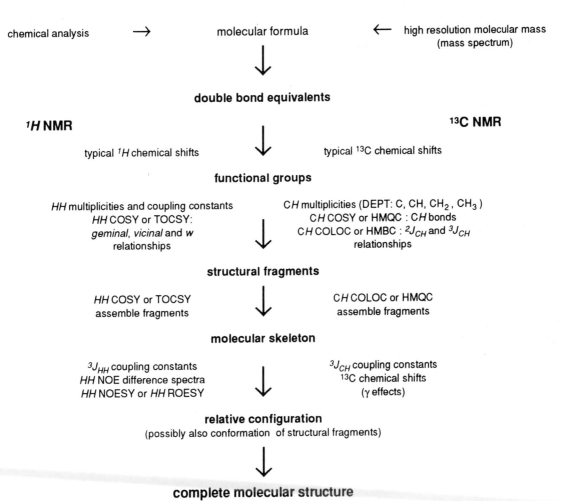

chemical analysis → molecular formula ← high resolution molecular mass (mass spectrum)

double bond equivalents

¹H **NMR**   **¹³C NMR**

typical *¹H* chemical shifts   typical ¹³C chemical shifts

functional groups

HH multiplicities and coupling constants   *CH* multiplicities (DEPT: C, CH, CH_2, CH_3)
HH COSY or TOCSY: *CH* COSY or HMQC : *CH* bonds
geminal, vicinal and *w*   *CH* COLOC or HMBC : $^2J_{CH}$ and $^3J_{CH}$
relationships   relationships

structural fragments

HH COSY or TOCSY   *CH* COLOC or HMQC
assemble fragments   assemble fragments

molecular skeleton

$^3J_{HH}$ coupling constants $^3J_{CH}$ coupling constants
HH NOE difference spectra   ¹³C chemical shifts
HH NOESY or *HH* ROESY   (γ effects)

relative configuration
(possibly also conformation of structural fragments)

complete molecular structure

If the amount of the sample is sufficient, then the carbon skeleton is best traced out from the two-dimensional INADEQUATE experiment. If the absolute configuration of particular C atoms is needed, the empirical applications of diastereotopism and chiral shift reagents are useful (Fig. 2.4). Anisotropic and ring current effects supply information about conformation (problem 34) and aromaticity (Section 2.5), and pH effects can indicate the site of protonation (problem 21). Temperature-dependent NMR spectra and ¹³C spin–lattice relaxation times (see Section 2.6) provide insight into molecular dynamics (problems 11 and 12).

3 PROBLEMS 1–50

In the following 50 problems, the chemical shift value (ppm) is given in the scale below the spectra and the coupling constant (Hz) is written immediately above or below the appropriate multiplet. Proton NMR data are italicised throughout in order to distinguish them from the parameters of other nuclei (^{13}C, ^{15}N).

1 *^{1}H* NMR spectrum **1** was obtained from dimethyl cyclopropanedicarboxylate. Is it a *cis* or a *trans* isomer?

Conditions: CDCl$_3$, 25 °C, 400 MHz.

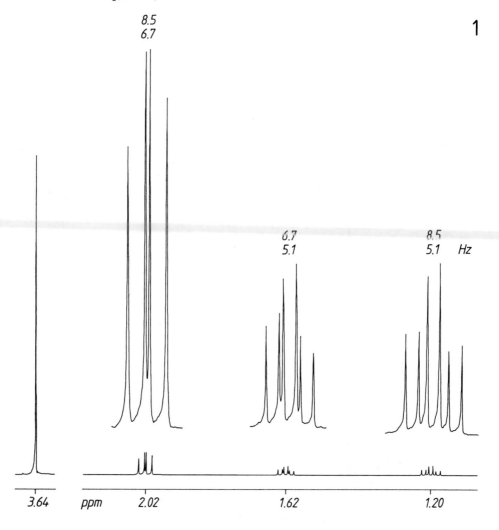

2   From which compound of formula $C_5H_8O_2$ was ^1H NMR spectrum **2** obtained?

Conditions: $CDCl_3$, 25 °C, 90 MHz.

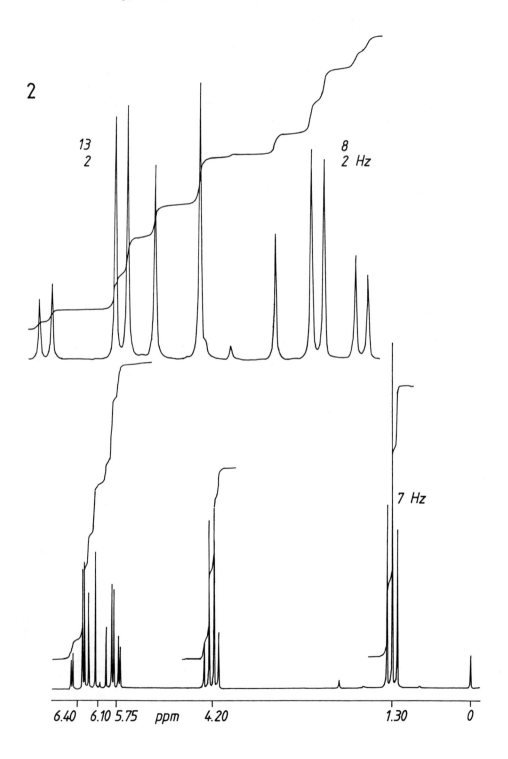

3 Which stereoisomer of the compound C_5H_6O is present given spectrum **3**?

Conditions: $CDCl_3$, 25 °C, 60 MHz (the only CW spectrum of this collection).

3

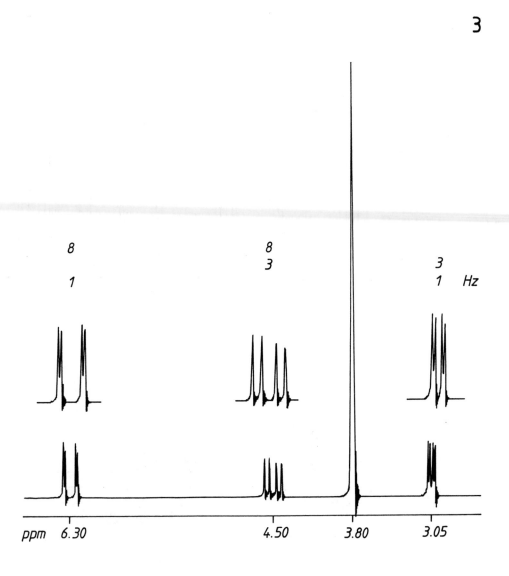

ppm 6.30   4.50   3.80   3.05

4   Which stereoisomer of the compound C_8H_9NO can be identified from ^1H NMR spectrum **4**?

Conditions: $CDCl_3$, 25 °C, 90 MHz.

4

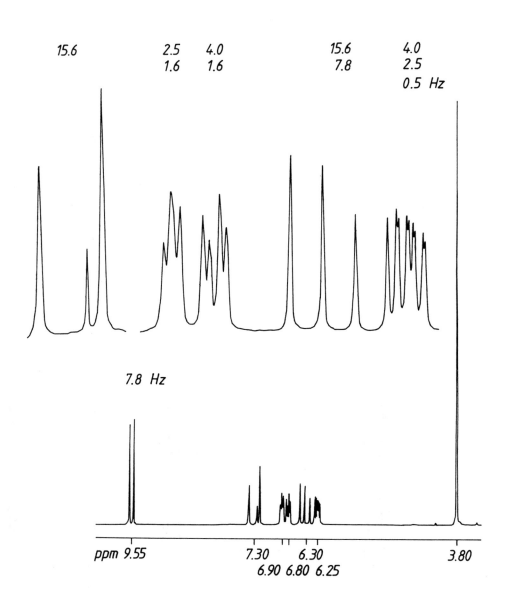

5 The reaction of 2,2'-bipyrrole with orthoformic acid triethyl ester in the presence of phosphoryl chloride ($POCl_3$) produced a compound which gave the ^1H NMR spectrum **5**. Which compound has been prepared?

Conditions: $CDCl_3$, 25 °C, 400 MHz.

Two broad D_2O-exchangeable signals at *11.6 ppm* (one proton) and *12.4 ppm* (two protons) are not shown.

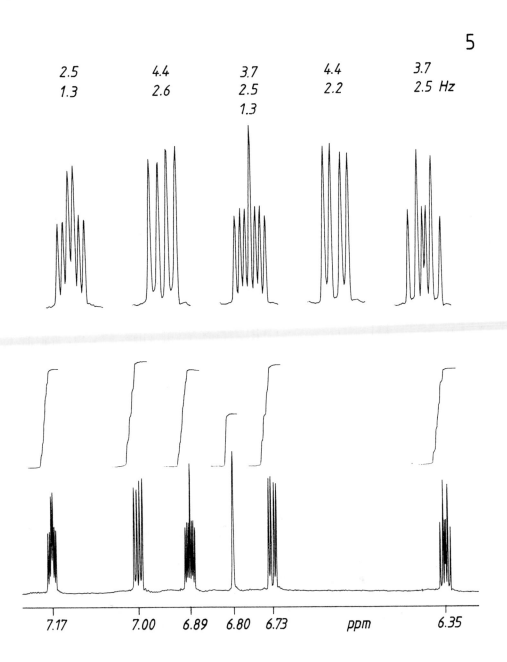

6   From which compound C_7H_7NO was ^1H NMR spectrum **6** obtained?

Conditions: $CDCl_3$, 25 °C, 90 MHz.

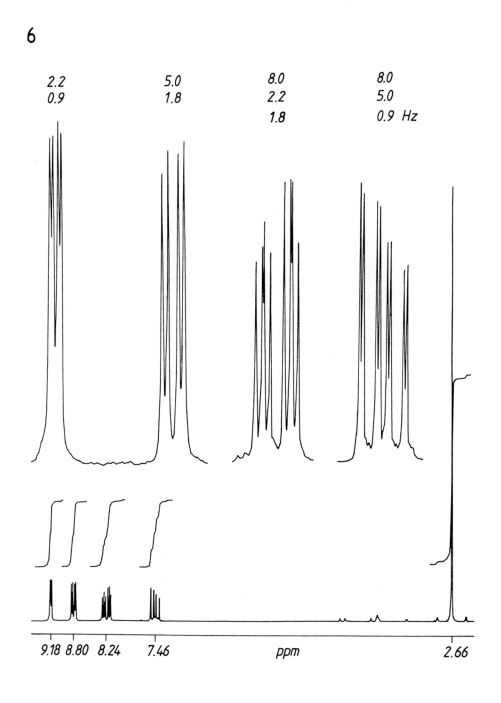

7   Which substituted isoflavone can be identified from ^1H NMR spectrum **7**?

Conditions: $CDCl_3$, 25 °C, 200 MHz.

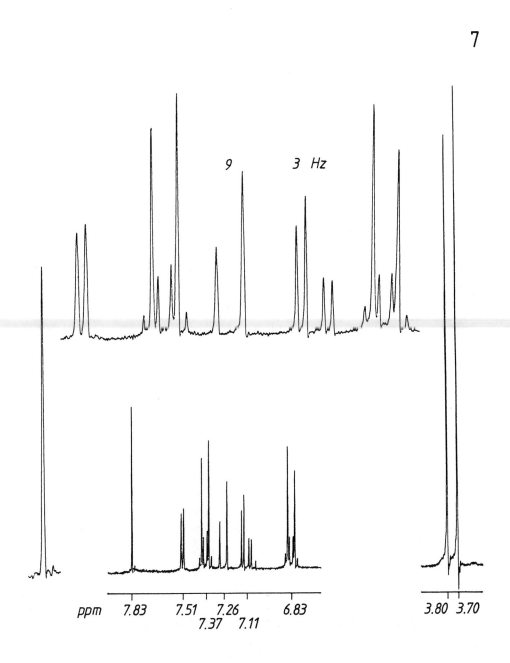

7

8 A natural substance of elemental composition $C_{15}H_{14}O_6$ was isolated from the plant *Centaurea chilensis* (Compositae). What is the structure and relative configuration of the substance given its ^1H NMR spectrum **8**?

Conditions: $CDCl_3$, 25 °C, 400 MHz.

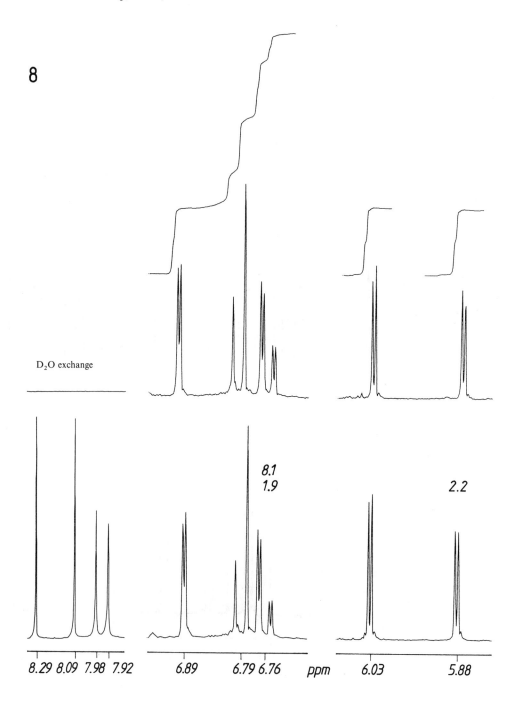

8.29 8.09 7.98 7.92   6.89   6.79 6.76   *ppm*   6.03   5.88

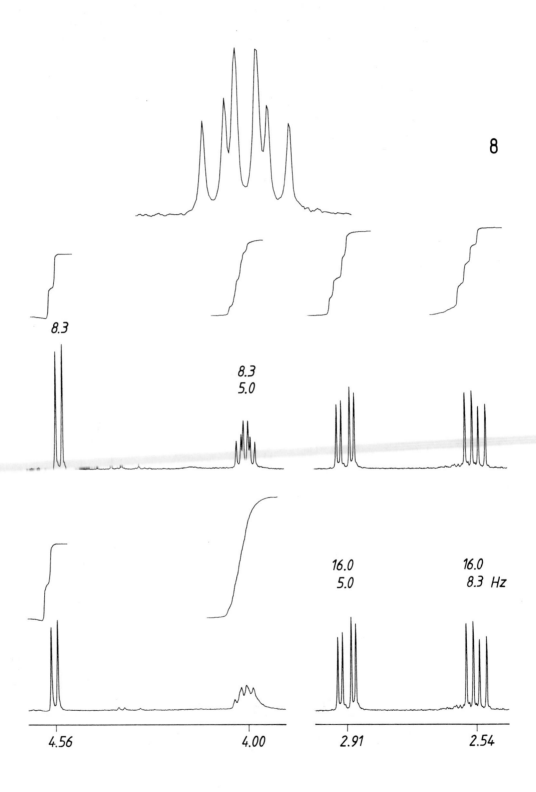

8

8.3

8.3
5.0

16.0
5.0

16.0
8.3 Hz

4.56   4.00   2.91   2.54

9 Characterisation of the antibiotic monordene with the elemental composition $C_{18}H_{17}O_6Cl$ isolated from *Monosporium bonorden* gave the macrolide structure **1**. The relative configuration of the *H* atoms on the two conjugated double bonds (6,7 *cis*, 8,9 *trans*) could be deduced from the 60 MHz ^1H NMR spectrum.[38] The relative configuration of the C atoms 2–5, which encompass the oxirane ring as a partial structure, has yet to be established.

The reference compound methyloxiráne gives the ^1H NMR spectrum **9a** shown with expanded multiplets. What information regarding its relative configuration can be deduced from the expanded ^1H multiplets of monordene displayed in **9b**?

Conditions: $(CD_3)_2CO$, 25 °C, 200 MHz.

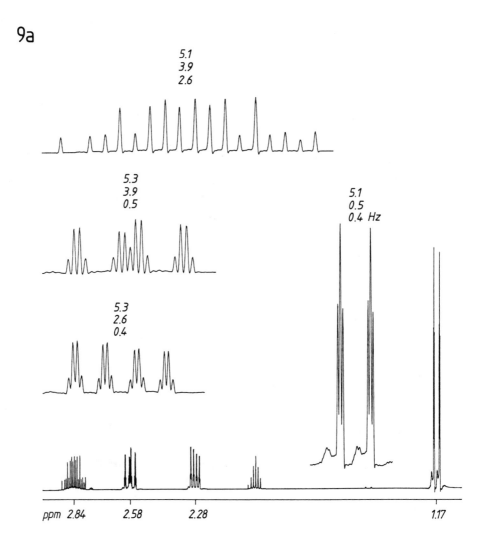

9a

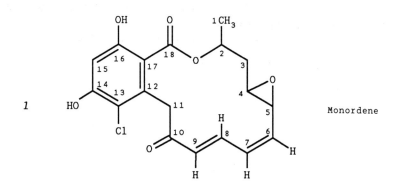

1   Monordene

9b

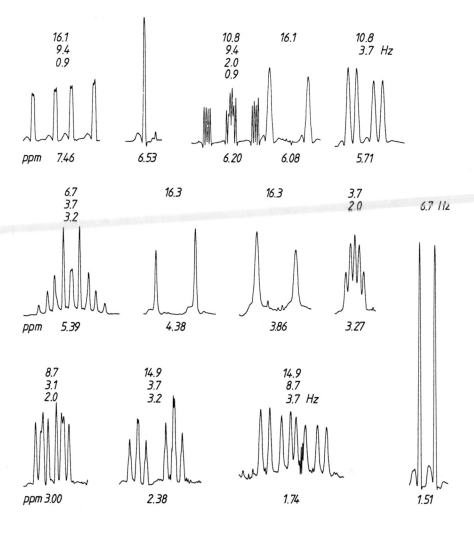

10 From the *HH* COSY contour plot **10a** it can be established which cycloadduct has been produced from 1-(*N*,*N*-dimethylamino)-2-methylbuta-1,3-diene and *trans*-β-nitrostyrene. The $^3J_{HH}$ coupling constant in the one-dimensional ^1H NMR spectrum **10b** can be used to deduce the relative configuration of the adduct.

Conditions: CDCl$_3$, 25 °C, 400 MHz.

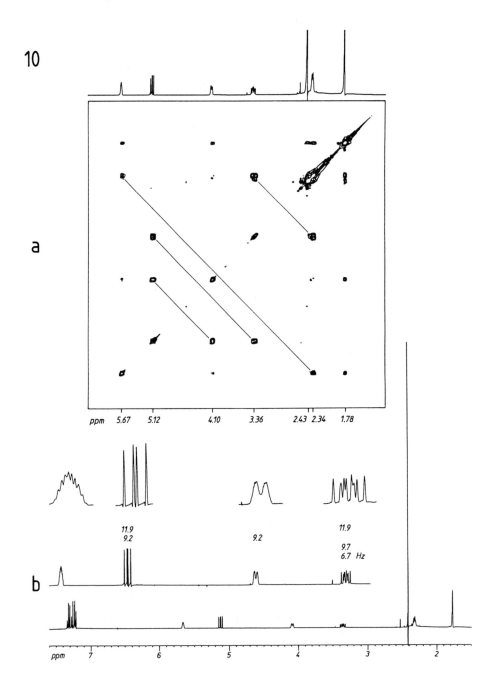

11 ^1H NMR spectra **11** were recorded from 3-(N,N-dimethylamino)acrolein at the temperatures given. What can be said about the structure of the compound and what thermodynamic data can be derived from these spectra?

Conditions: CDCl$_3$, 50% v/v, 250 MHz.

11

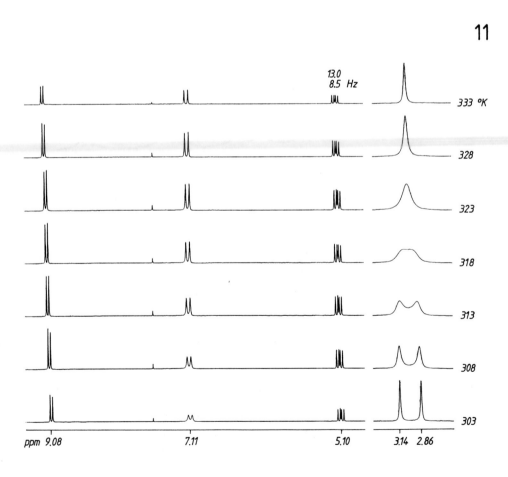

ppm 9.08   7.11   5.10   3.14   2.86

12 ^{13}C NMR spectra **12** were recorded using *cis*-1,2-dimethylcyclohexane at the temperatures given; the DEPT experiment at 223 K was also recorded in order to distinguish the C*H* multiplicities (C*H* and C*H$_3$* positive, C*H$_2$* negative). Which assignments of resonances and what thermodynamic data can be deduced from these spectra?

Conditions: $(CD_3)_2CO$, 95% v/v, 100 MHz, ^1H broadband decoupled.

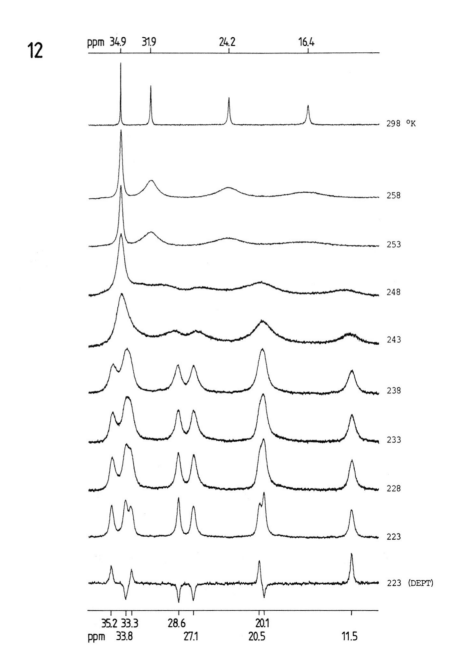

13 No further information is required to deduce the identity of this compound from ^{13}C NMR spectra **13**.

Conditions: CDCl$_3$, 25 °C, 20 MHz. (a) Proton broadband decoupled spectrum; (b) NOE enhanced coupled spectrum (gated decoupling); (c) expanded section of (b).

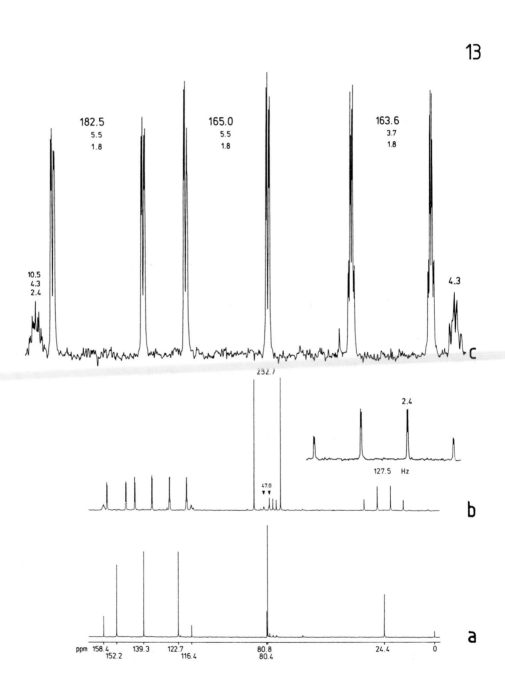

14 In hexadeuteriodimethyl sulphoxide the compound which is labelled as 3-methyl-pyrazolone gives ^{13}C NMR spectra **14**. In what form is this compound present in this solution?

Conditions: $(CD_3)_2SO$, 25 °C, 20 MHz. (a) ^1H broadband decoupled spectrum; (b) NOE enhanced coupled spectrum (gated decoupling); (c) expanded sections of (b).

14

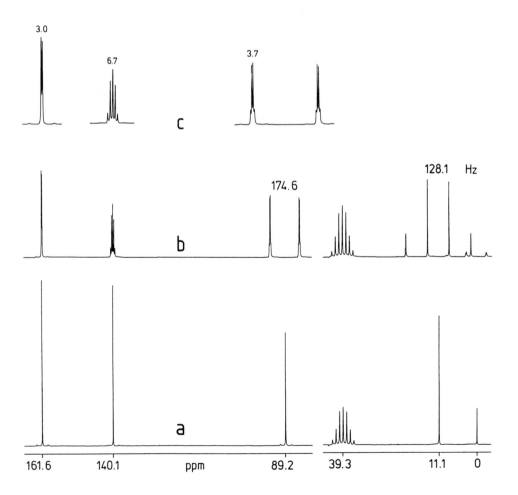

15   Using which compound $C_8H_8O_2$ were the ^{13}C NMR spectra **15** recorded?

Conditions: CDCl$_3$: (CD$_3$)$_2$CO (1:1), 25 °C, 20 MHz. (a) ^1H broadband decoupled spectrum; (b) NOE enhanced coupled spectrum (gated decoupling); (c) expanded section of (b).

15

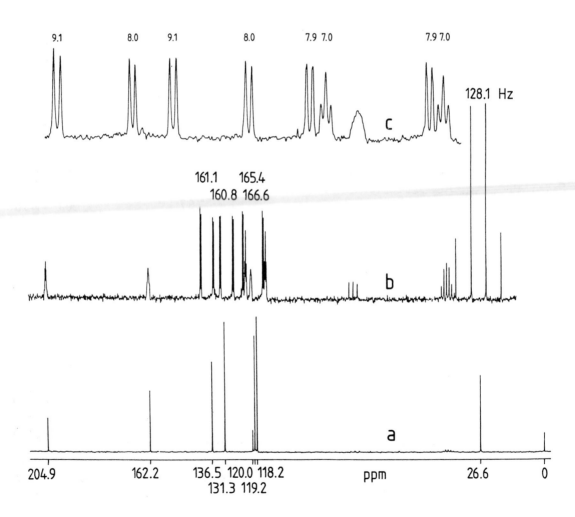

16 1,3,5-Trinitrobenzene reacts with dry acetone in the presence of potassium methoxide to give a crystalline violet compound $C_9H_8N_3O_7K$. Deduce its identity from the ^{13}C NMR spectra **16**.

Conditions: $(CD_3)_2SO$, 25 °C, 22.63 MHz. (a) ^1H broadband decoupled spectrum; (b) without decoupling; (c) expanded section of (b).

16

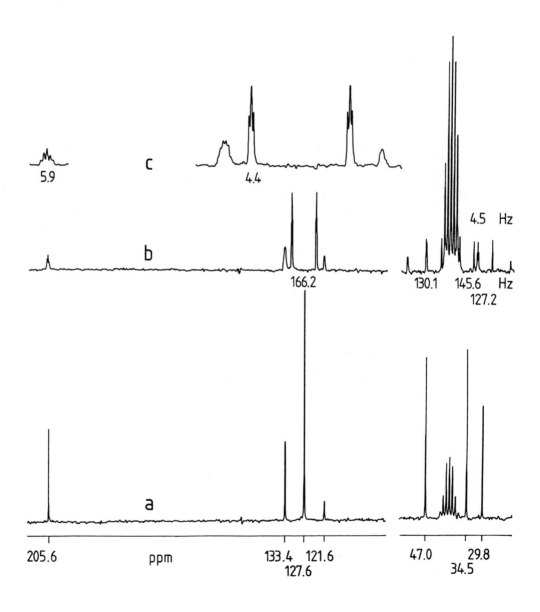

17 3-[4-(*N,N*-Dimethylamino)phenyl]-2-ethylpropenal (*3*) was produced by reaction of *N,N*-dimethylaniline (*1*) with 2-ethyl-3-ethoxyacrolein (*2*) in the presence of phosphorus oxytrichloride.

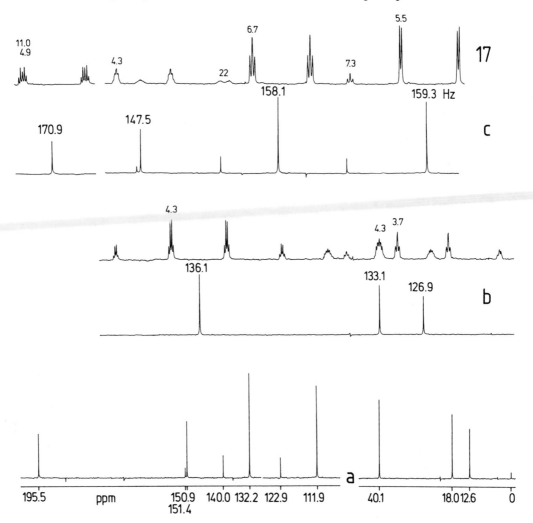

Since the olefinic CC double bond is trisubstituted, the relative configuration cannot be determined on the basis of the *cis* and *trans* couplings of *vicinal* alkene protons in the ^1H NMR spectrum. What is the relative configuration given the ^{13}C NMR spectra **17**?

Conditions: CDCl$_3$, 25 °C, 20 MHz. (a) ^1H broadband decoupled spectrum; (b) expanded sp^3 shift range; (c) expanded sp^2 shift range; (b) and (c) each with the ^1H broadband decoupled spectrum below and NOE enhanced coupled spectrum above.

18 2-Trimethylsilyloxy-β-nitrostyrene was the target of Knoevenagel condensation of 2-trimethylsiloxybenzaldehyde with nitromethane in the presence of *n*-butylamine as base. ^{13}C NMR spectra **18** were obtained from the product of the reaction. What has happened?

Conditions: CDCl$_3$, 25 °C, 20 MHz. (a) sp^3 shift range; (b) sp^2 shift range, in each case with the ${}^1H$ broadband decoupled spectrum below and the NOE enhanced coupled spectrum (gated decoupling).

18

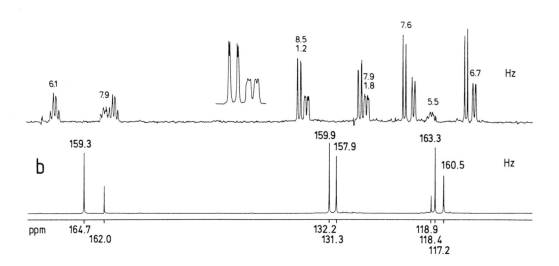

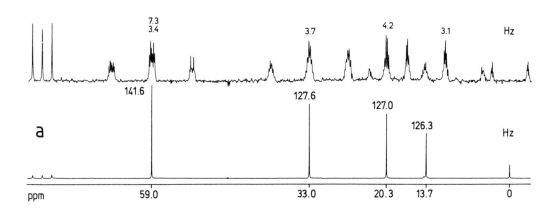

19 From which compound were the INADEQUATE contour plot and ^{13}C NMR spectra **19** obtained?

Conditions: $(CD_3)_2CO$, 95% v/v, 25 °C, 100 MHz. (a) Symmetrised INADEQUATE contour plot with ^{13}C NMR spectra; (b) ^1H broadband decoupled spectrum; (c) NOE enhanced coupled spectrum (gated decoupling); (d) expansion of multiplets of (c).

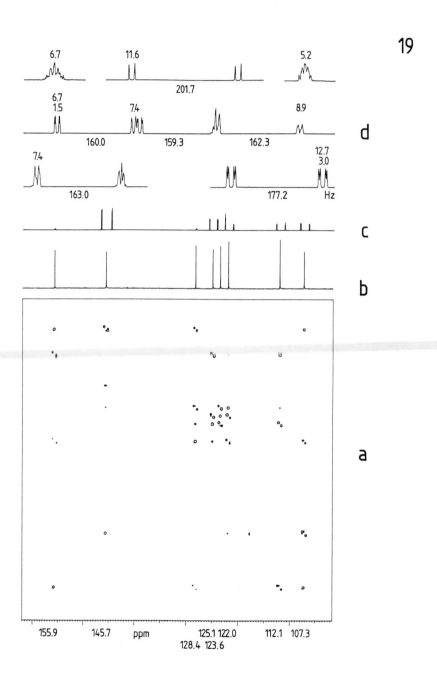

20 The hydrolysis of 3-ethoxy-4-ethylbicyclo[4.1.0]hept-4-en-7-one propylene acetal (*1*) with aqueous acetic acid in tetrahydrofuran gives an oil with the molecular formula $C_{12}H_{18}O_3$, from which the INADEQUATE contour plot **20** and DEPT spectra were obtained. What is the compound?

Conditions: $(CD_3)_2CO$, 95% v/v, 25 °C, 100 MHz. (a) Symmetrised INADEQUATE contour plot; (b–d) ^{13}C NMR spectra, (b) ^1H broadband decoupled spectrum, (c) DEPT CH subspectrum, (d) DEPT spectrum, CH and CH_3 positive and CH_2 negative.

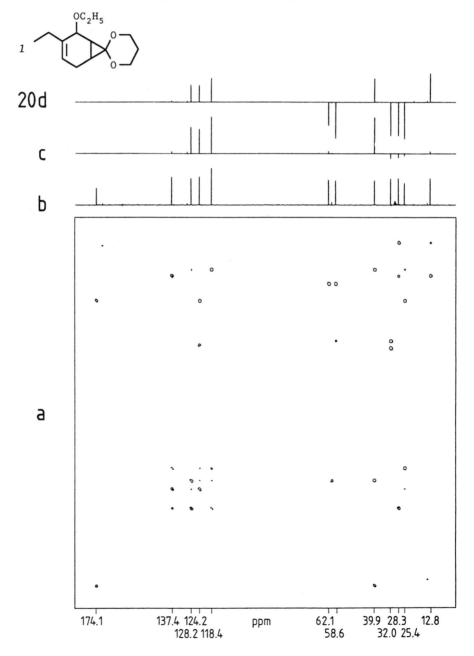

21   Procaine hydrochloride gives the ^1H and ^{13}C NMR spectra **21**. Which amino group is protonated?

Conditions: $CDCl_3$: $(CD_3)_2SO$ (3:1), 25 °C, 20 MHz (^{13}C), 200 MHz (^1H). (a) ^1H spectrum with expanded multiplets; (b–d) ^{13}C NMR spectra; (b) ^1H broadband decoupled spectrum; (c) NOE enhanced coupled spectrum (gated decoupling); (d) expansion of multiplets (113.1–153.7 ppm).

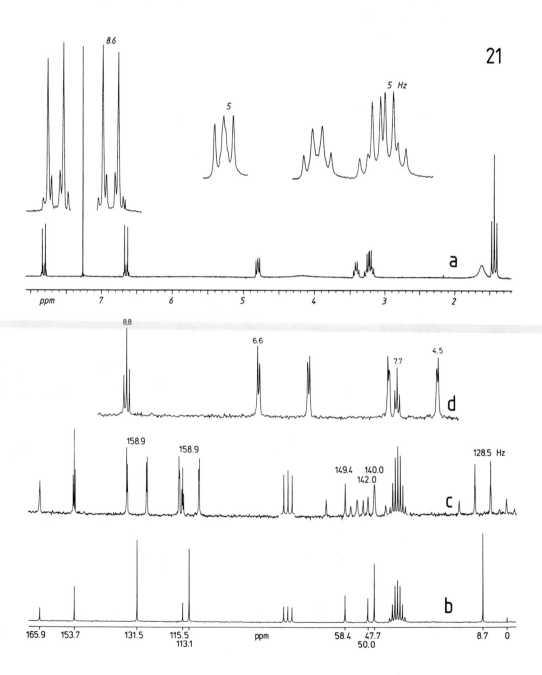

22   3,4-Dihydro-2*H*-pyran-5-carbaldehyde (*1*) was treated with sarcosine ethyl ester (*2*) in the presence of sodium ethoxide. What is the structure of the crystalline product $C_{11}H_{16}NO_3$ given its NMR spectra **22**?

Conditions: CDCl$_3$, 25 °C, 100 MHz (^{13}C), 400 MHz (^{1}H). (a) ^{1}H NMR spectrum with expanded multiplets; (b–e) ^{13}C NMR partial spectra; (b) sp^3 shift range; (c) sp^2 shift range, each with ^{1}H broadband decoupled spectrum (bottom) and the NOE enhanced coupled spectrum (gated decoupling, top) with expansions of multiplets (d) (59.5–61.7 ppm) and (e) (117.0– 127.5 ppm).

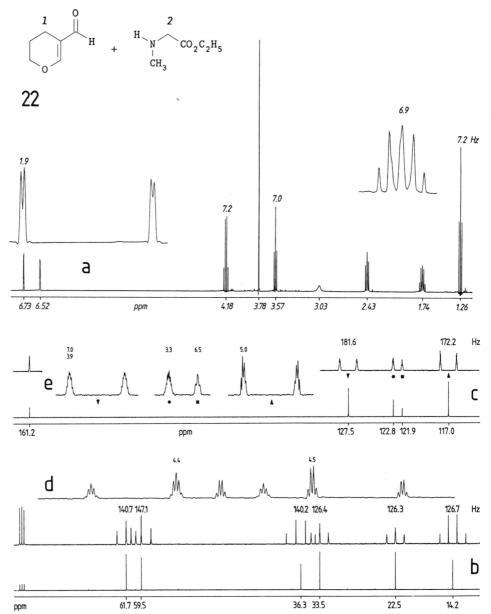

23 1-Ethoxy-2-propylbuta-1,3-diene and *p*-tolyl sulphonyl cyanide react to give a crystalline product. What is this product given its NMR spectra set **23**?

Conditions: CDCl₃, 25 °C, 100 MHz (¹³C), 400 MHz (¹H). (a–e) ¹³C NMR spectra; (a,b) ¹H broadband decoupled spectra; (c,d) NOE enhanced coupled spectra (gated decoupling) with expansion (e) of the multiplets in the sp² shift range; (f) ¹H NMR spectrum with expanded multiplets.

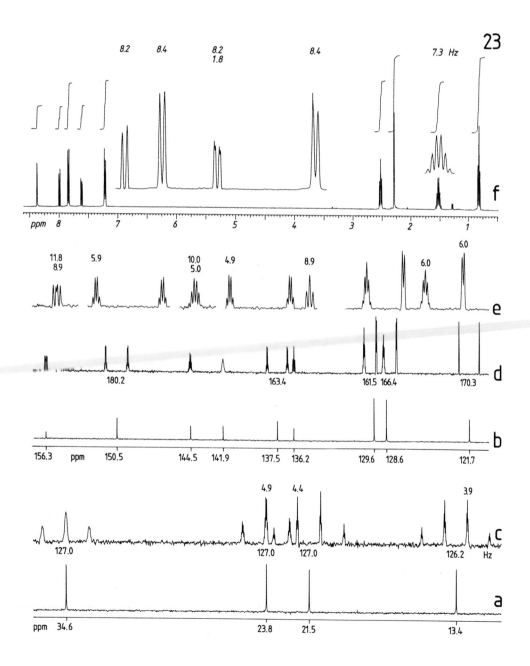

24   5-Amino-1,2,4-triazole undergoes a cyclocondensation with 3-ethoxyacrolein (*1*) to form 1,2,4-triazolo[1,5-*a*]pyrimidine (*3*) or its [4,3-*a*] isomer (*5*), according to whether it reacts as *1H* or *4H* tautomer *2* or *4*. Moreover, the pyrimidines *3* and *5* can interconvert by a Dimroth rearrangement. Since the ^1H NMR spectrum **24a** does not enable a clear distinction to be made (*AMX* systems for both pyrimidine protons in both isomers), the ^{13}C and ^{15}N NMR spectra **24b–d** were obtained. What is the compound?

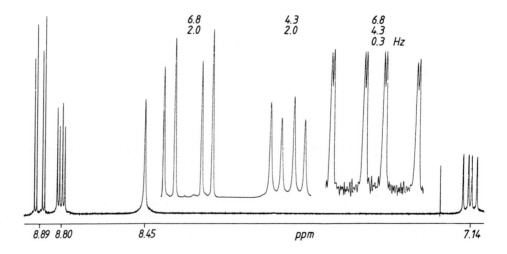

Conditions: CDCl$_3$ (a–c), (CD$_3$)$_2$SO (d), 25 °C, 200 MHz (^1H), 20 MHz (^{13}C), 40.55 MHz (^{15}N). (a) ^1H NMR spectrum; (b, c) ^{13}C NMR spectra; (b) ^1H broadband decoupled spectrum; (c) NOE enhanced coupled spectrum (gated decoupling); (d) ^{15}N NMR spectrum, without decoupling, with expanded multiplets, ^{15}N shifts calibrated relative to ammonia as reference.[7, 8]

24a

6.8
2.0

4.3
2.0

6.8
4.3
0.3 Hz

8.89 8.80   8.45   *ppm*   7.14

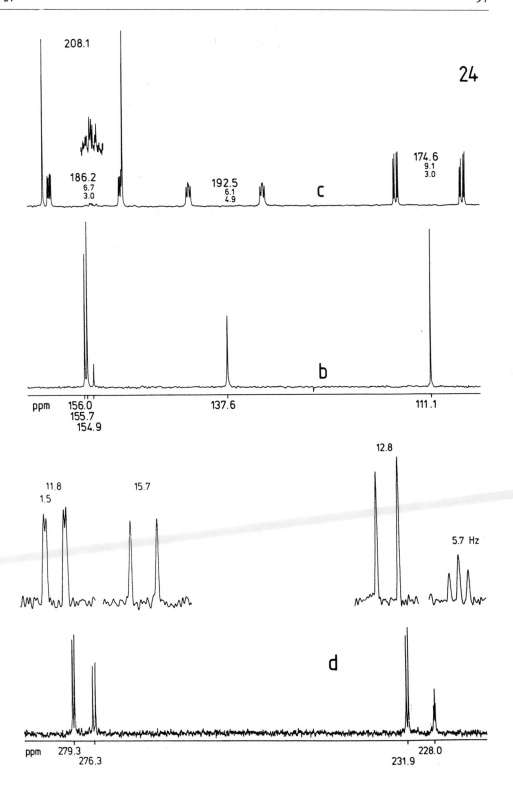

25 NMR spectra **25** were obtained from 6-butyltetrazolo[1,5-*a*]pyrimidine (*1*). What form does the heterocycle take?

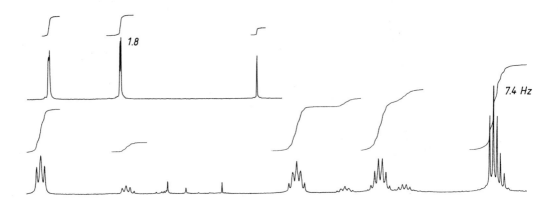

1

Conditions: $(CD_3)_2CO$, 25 °C, 400 MHz (^1H), 100 MHz (^{13}C), 40.55 MHz (^{15}N). (a) ^1H NMR spectrum with expanded partial spectra and integrals; (b, c) ^{13}C NMR spectra, in each case showing proton broadband decoupled spectrum below and gated decoupled spectrum above, (b) aliphatic resonances and (c) heteroaromatic resonances; (d) ^{15}N NMR spectrum, coupled, with expanded sections and integrals.

25a

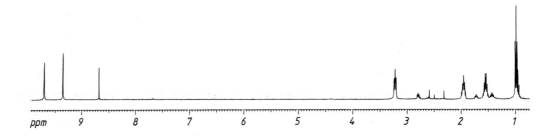

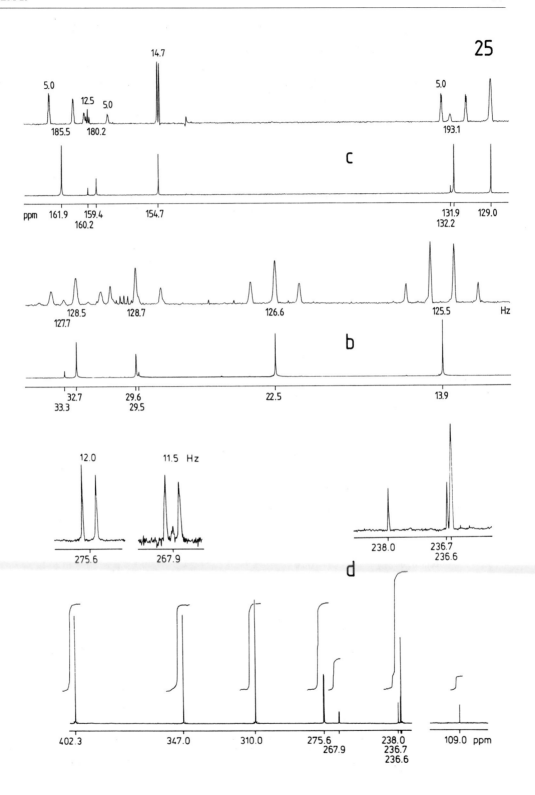

26 From which compound $C_6H_{10}O$ were the NMR spectra **26** recorded?

Conditions: $(CD_3)_2CO$, 25 °C, 200 MHz (^1H), 50 MHz (^{13}C). (a) ^1H NMR spectrum with expanded sections; (b, c) ^{13}C NMR partial spectra, each with proton broadband decoupled spectrum below and NOE enhanced coupled spectrum above with expanded multiplets at 76.6 and 83.0 ppm; (d) symmetrised INADEQUATE contour plot.

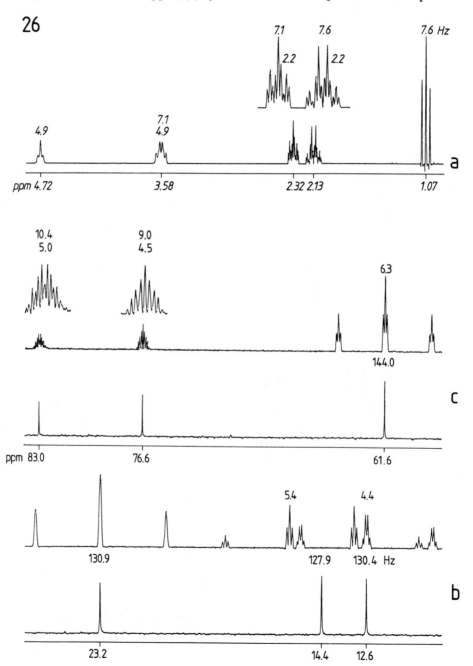

26d

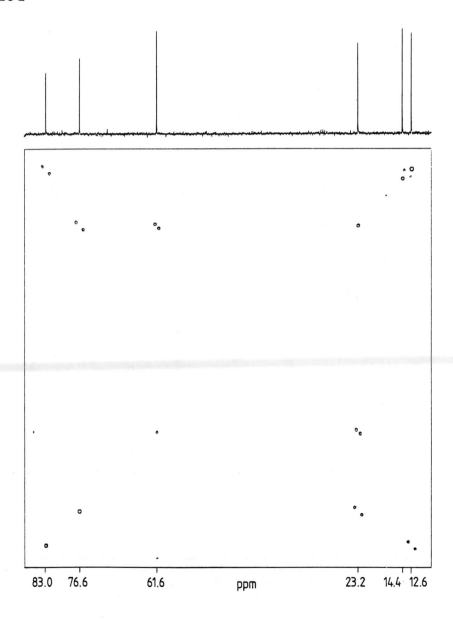

27 Which compound of formula $C_{11}H_{22}O_2$ gives the NMR spectra set **27**?

Conditions: $CDCl_3$, 25 °C, 200 MHz (^1H), 50 MHz (^{13}C). (a) ^1H NMR spectrum with expansion (b) and NOE difference spectra (c, d), with decoupling at *2.56 ppm* (c) and *2.87 ppm* (d); (e–g) ^{13}C NMR spectra; (e) ^1H broadband decoupled spectrum; (f) NOE enhanced coupled spectrum (gated decoupling) with expansions (g) (23.6, 113.3, 113.8, 127.0, 147.8, 164.6 and 197.8 ppm); (h) section of the *CH* COSY diagram.

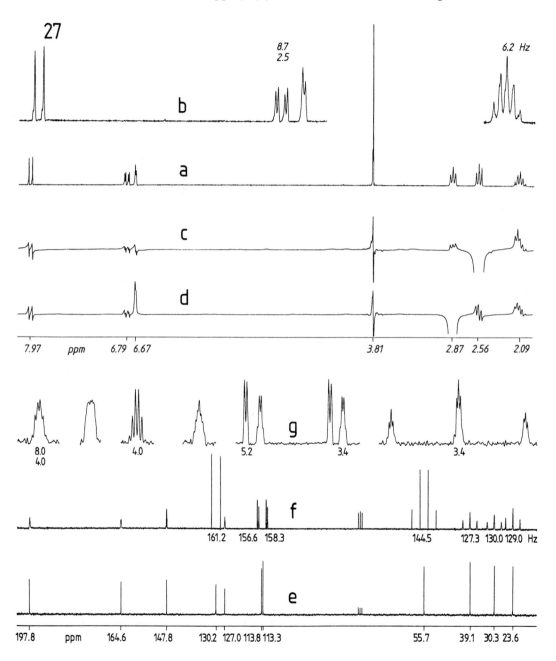

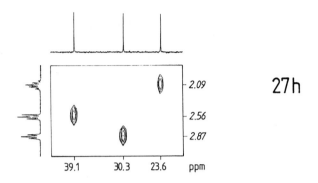

27h

28 Phthalaldehydic acid (*o*-formylbenzoic acid) gives the NMR spectra set **28**. In what form does the compound actually exist?

Conditions: CDCl$_3$: (CD$_3$)$_2$SO (9:1), 25 °C, 80 and 400 MHz (^1H), 20 and 100 MHz (^{13}C). (a) ^1H NMR spectrum with expanded section (b) before and after D$_2$O exchange; (c, d) ^{13}C NMR spectra; (c) ^1H broadband decoupled spectrum; (d) NOE enhanced coupled spectrum (gated decoupling); (e) CH COSY diagram (100/400 MHz); (f) HH COSY diagram (400 MHz). The ordinate scales in (e) and (f) are the same.

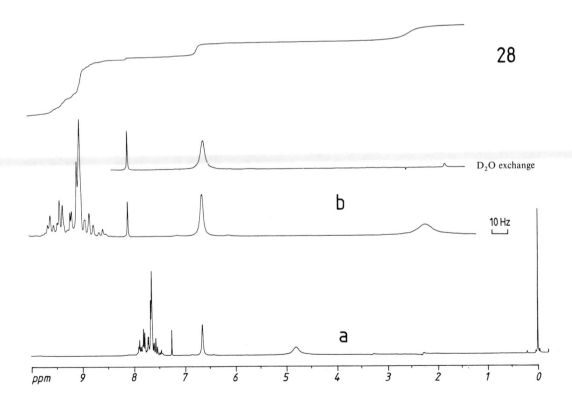

28

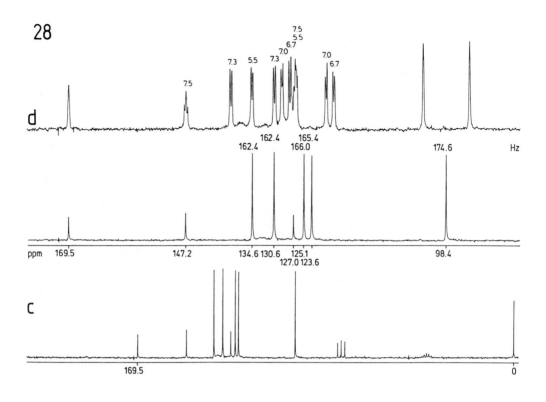

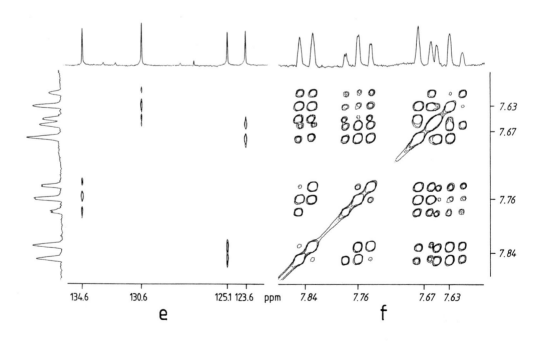

e

f

29   A fragrant substance found in cucumber and melon produces the NMR spectra set **29**. The identity and structure of the substance can be derived from these spectra without any further information.

Conditions: CDCl$_3$, 30 mg per 0.3 ml, 25 °C, 400 MHz (^1H), 100 MHz (^{13}C. (a) *HH* COSY diagram; (b) ^1H NMR spectrum with expanded multiplets; (c) ^{13}C NMR partial spectra, each with ^1H broadband decoupled spectrum below and NOE enhanced coupled spectrum (gated decoupling) above; (d) C*H* COSY diagram with expanded section (133.2– 133.3 ppm).

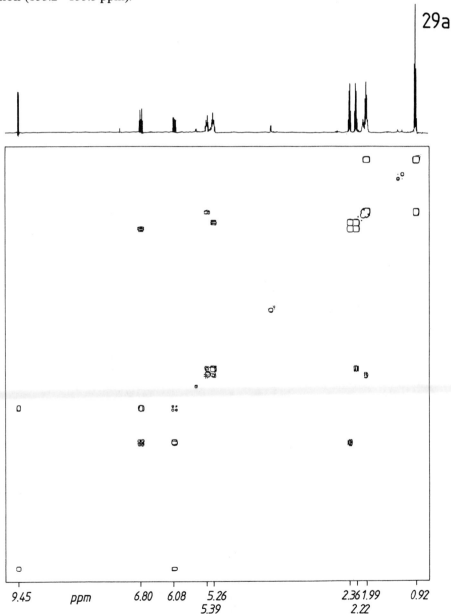

29a

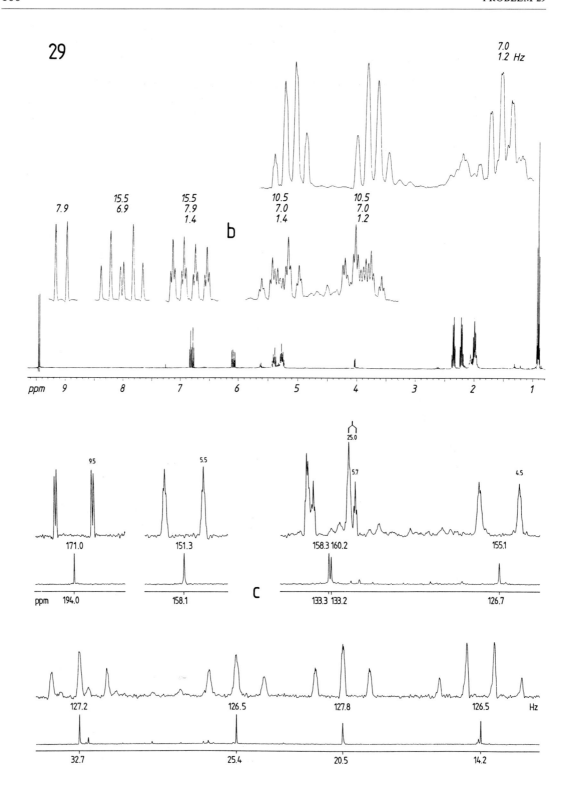

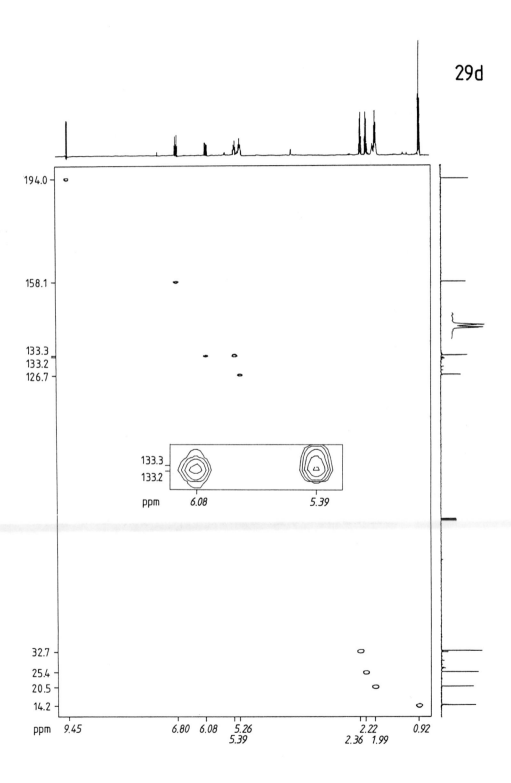

29d

30

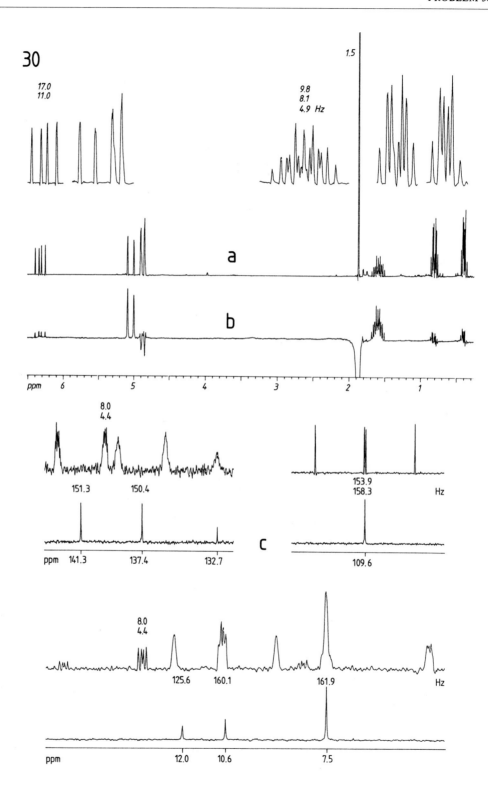

30 Several shifts and coupling constants in the NMR spectra set **30** are so typical that the carbon skeleton can be deduced without any additional information. An NOE difference spectrum gives the relative configuration of the compound.

Conditions: CDCl₃, 25 °C, 200 MHz (¹H), 50 MHz (¹³C). (a) ¹H NMR spectrum with expanded multiplets; (b) NOE difference spectrum, irradiated at *1.87 ppm*; (c) ¹³C NMR partial spectra, each with ¹H broadband decoupled spectrum below and NOE enhanced coupled spectrum (gated decoupling) above; (d) CH COSY diagram ('empty' shift ranges omitted).

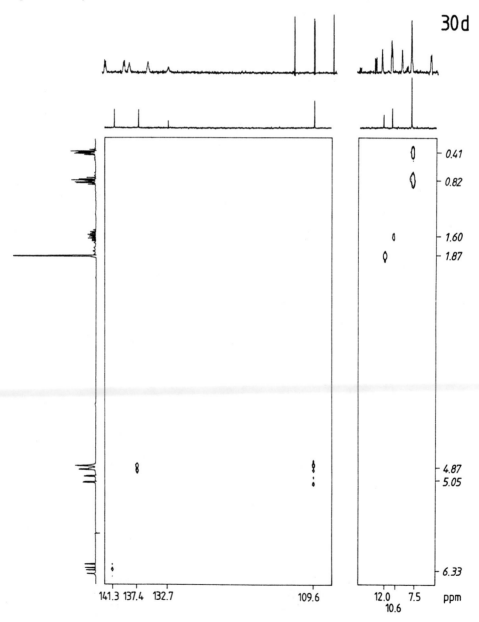

30d

31 Commercial cyclopentadiene produces the NMR spectra **31**. In what form does this compound actually exist, and what is its relative configuration?

Conditions: $(CD_3)_2CO$, 95% v/v, 25 °C, 400 and 200 MHz (^1H), 100 MHz (^{13}C). (a) ^1H NMR spectrum with expansion (above) and *HH* COSY diagram (below); (b) NOE difference spectra (200 MHz) with decoupling at *1.25* and *1.47 ppm*; (c, d) *CH* COSY contour plots with DEPT subspectra to distinguish *CH* (positive) and *CH₂* (negative); (c) alkyl shift range; (d) alkenyl shift range.

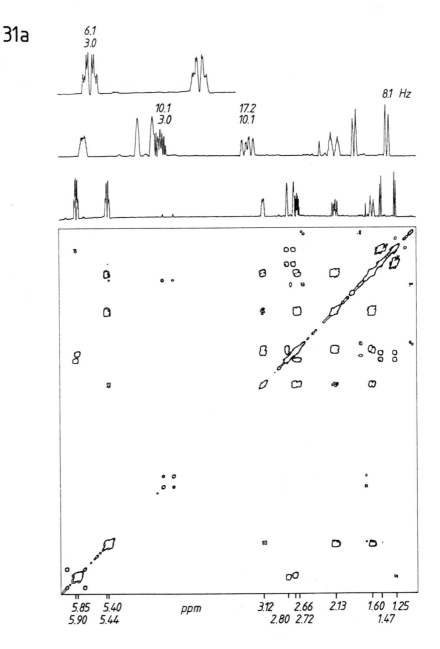

31a

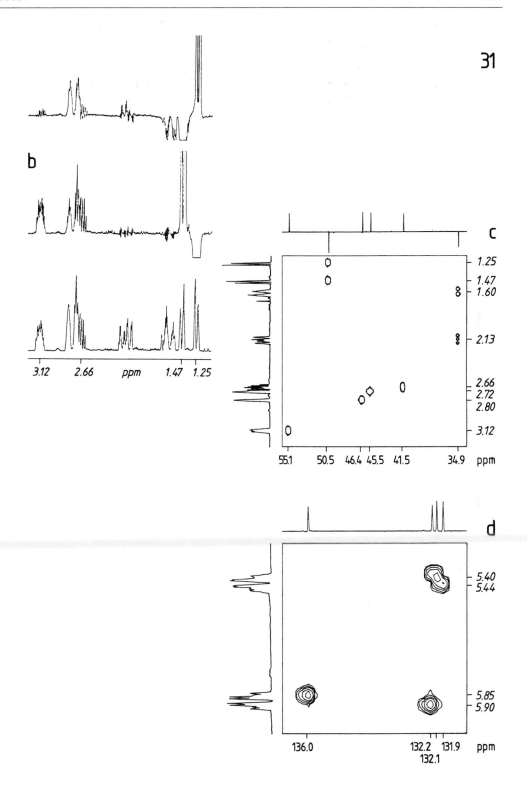

32 Which C*H* skeleton can be deduced from the NMR experiments **32**? What relative configuration does the *¹H* NMR spectrum indicate?

Conditions: (CD₃)₂CO, 90% v/v, 25 °C, 400 MHz (*¹H*), 50 and 100 MHz (*¹³C*). (a) Symmetrised INADEQUATE diagram (50 MHz); (b) C*H* COSY diagram with expansion (c) of the *¹H* NMR spectrum between *1.5* and *2.3 ppm*.

32a

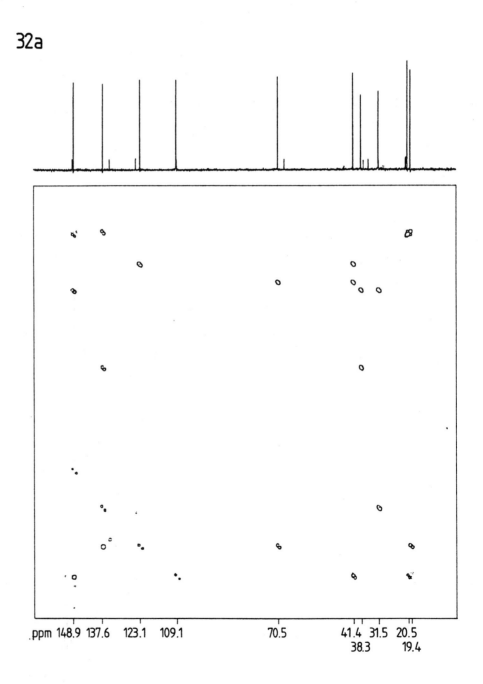

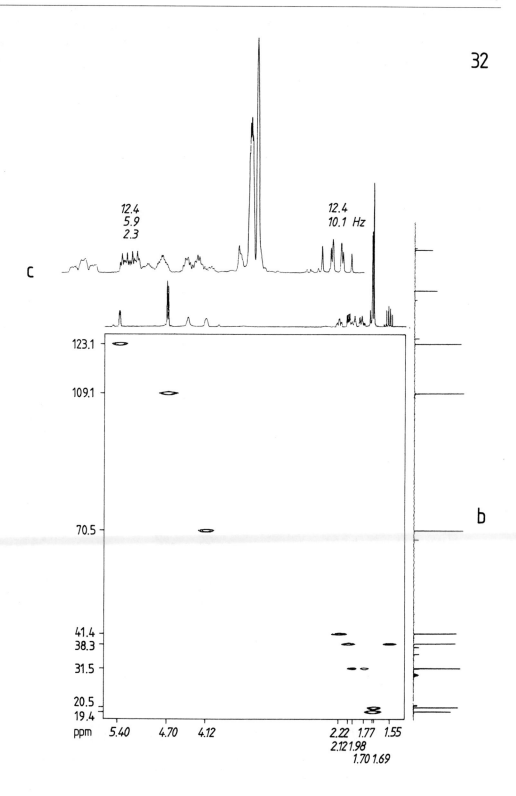

32

12.4
5.9
2.3

12.4
10.1 Hz

c

b

123.1

109.1

70.5

41.4
38.3

31.5

20.5
19.4

ppm   5.40   4.70   4.12   2.22 1.77 1.55
                                               2.12 1.98
                                                 1.70 1.69

33 Menthane-3-carboxylic acid (*4*) is synthesised from (−)-menthol (*1*) via menthyl chloride (*2*) and its Grignard reagent (*3*). The product *4* with melting point 56–60 °C and specific rotation $[\alpha]_D^{20} = -41.2°$ (ethanol, $c = 8.16$) gives the NMR results **33**. What is its configuration and what assignments of the signals can be made given these NMR experiments?

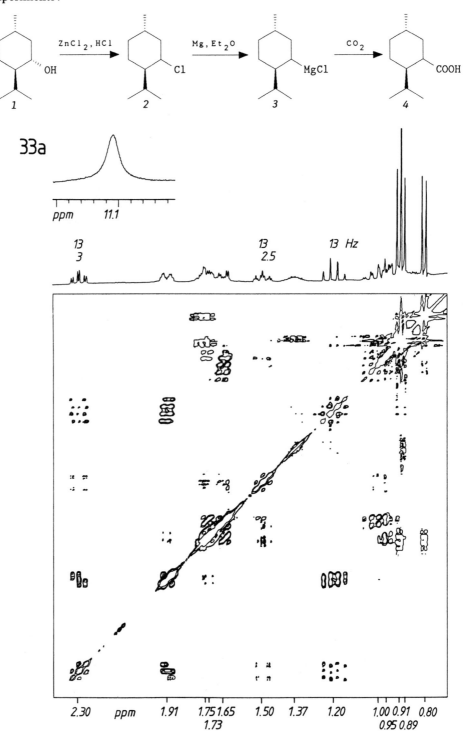

Conditions: CDCl$_3$, 25 °C, 400 MHz (^1H), 100 MHz (^{13}C). (a) ^1H NMR spectrum and *HH* COSY plot; (b) C*H* COSY diagram with DEPT C*H* subspectrum (c) and DEPT spectrum (d), C*H* and C*H$_3$* positive, C*H$_2$* negative.

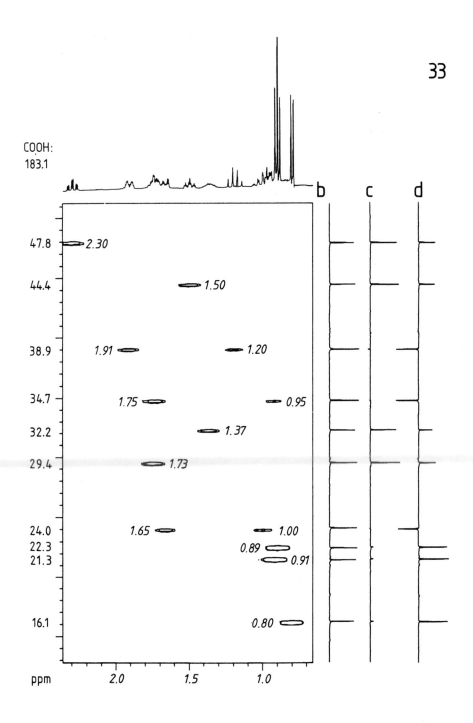

COOH:
183.1

34 *meso*-α,α,α,α-Tetrakis{2[(*p*-menth-3-ylcarbonyl)amino]phenyl}porphyrin is prepared by acylation of *meso*-α,α,α,α-tetrakis(2-aminophenyl)porphyrin with (+)- or
(−)-3-menthanecarboxylic acid chloride. What spatial arrangement of the menthyl
residue is indicated by the ^1H shifts of the chiral porphyrin framework?

Conditions: CDCl$_3$, 25 °C, 400 MHz (^1H), 100 MHz (^{13}C). (a) ^1H NMR spectrum; (b)
^{13}C NMR spectrum, aliphatic region below, aromatic region above; (c) CH COSY
diagram of aliphatic shift range with DEPT subspectrum (CH and CH_3 positive, CH_2
negative).

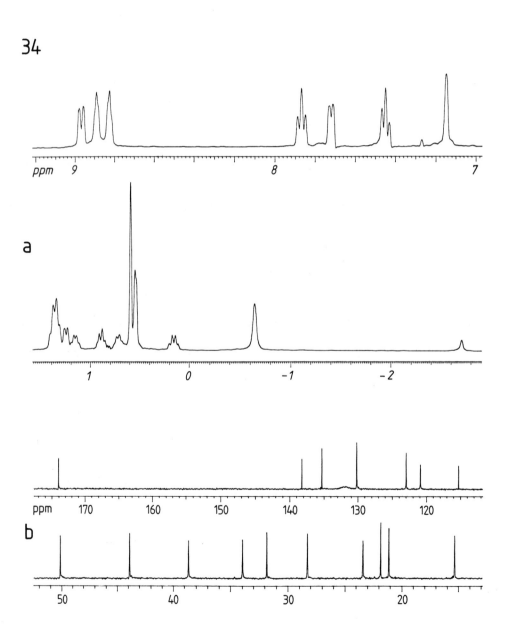

34c

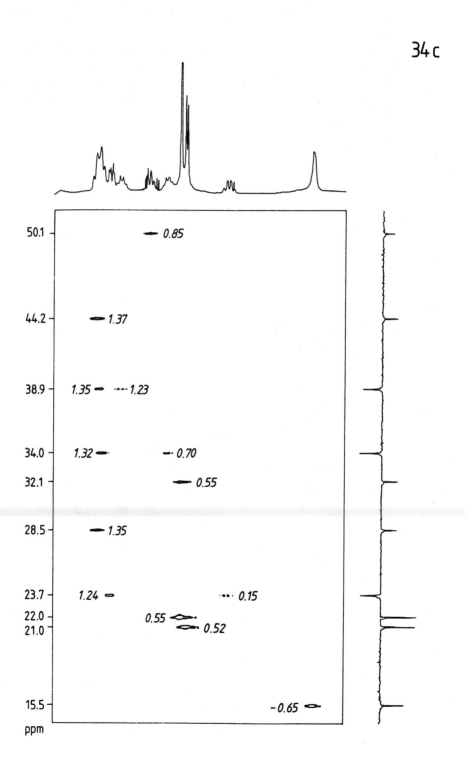

35   Cyclohexene oxide and metallated 2-methylpyridine reacted to give a product which gave the NMR results **35**. Identify the relative configuration of the product and assign the resonances.

Conditions: CDCl₃, 25 °C, 400 MHz (¹H), 100 MHz (¹³C). (a, b) ¹H NMR spectra, aromatic region (a), aliphatic region (b); (c) *HH* COSY plot of aliphatic shift range; (d) *CH* COSY plot with DEPT subspectra to distinguish *CH* and *CH₂*; (e) *CH* COSY diagram showing the region from 24.7 to 45.4 ppm; (f) symmetrised INADEQUATE plot of aliphatic region.

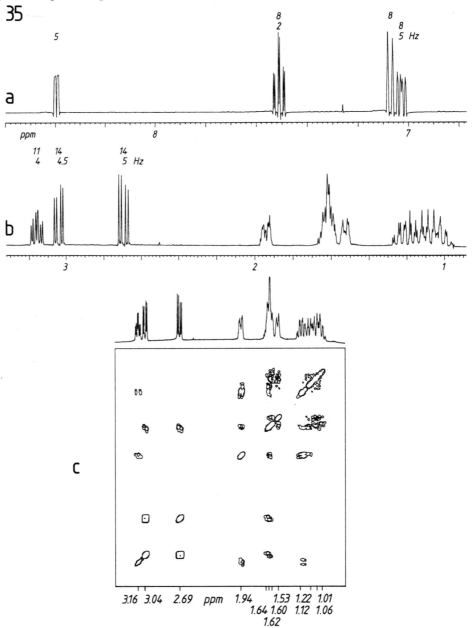

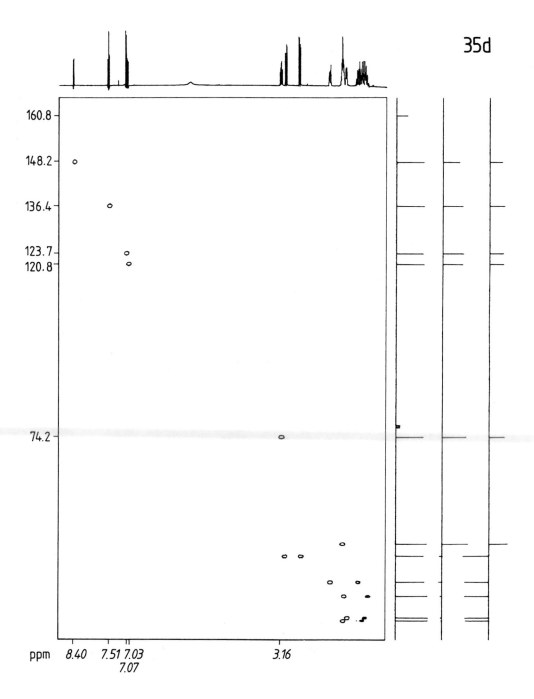

35d

ppm   8.40   7.51 7.03   3.16
             7.07

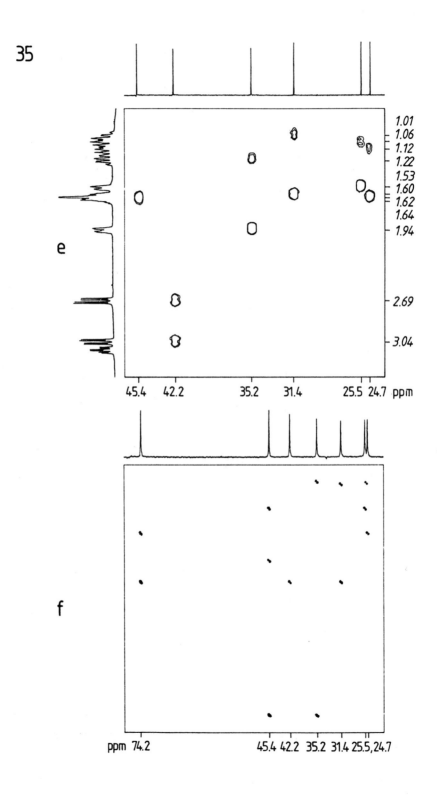

36   What compound $C_{18}H_{20}O_6$ can be identified from the *CH* COSY and *CH* COLOC diagrams **36** and the ^1H NMR spectra shown above?

Conditions: $(CD_3)_2SO$, 25 °C, 200 MHz (^1H), 50 MHz (^{13}C). *CH* COSY (shaded contours) and *CH* COLOC plot (unshaded contours) in one diagram; in the ^1H NMR spectrum the signal at *12.34 ppm* disappears following D_2O exchange.

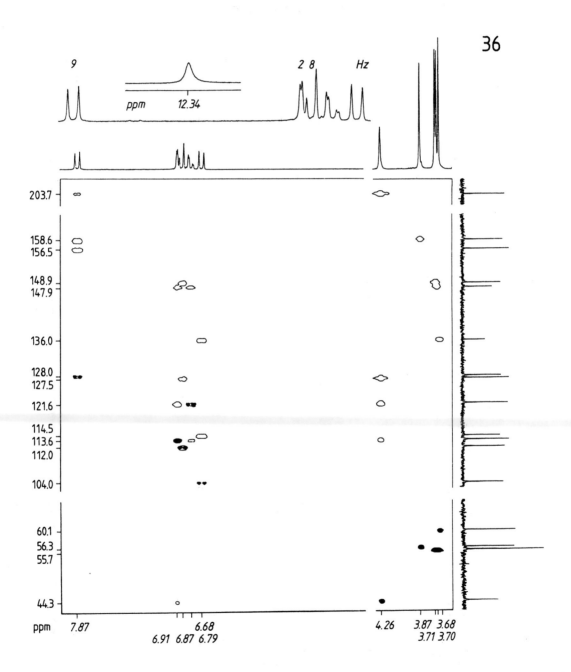

37   What compound $C_{19}H_{18}O_6$ can be identified from the *CH* COSY and *CH* COLOC plots **37** and from the *^1H* NMR spectra shown above?

Conditions: $(CD_3)_2SO$, 25 °C, 200 MHz (*^1H*), 50 MHz (*^{13}C*). (a) *CH* COSY diagram (shaded contours) and *CH* COLOC spectra (unshaded contours) in one diagram with an expansion of the *^1H* NMR spectrum; (b) part of the aromatic shift range of (a); (c) parts of *^{13}C* NMR spectra to be assigned, with *^1H* broadband decoupled spectrum below and NOE enhanced coupled spectrum (gated decoupling) above.

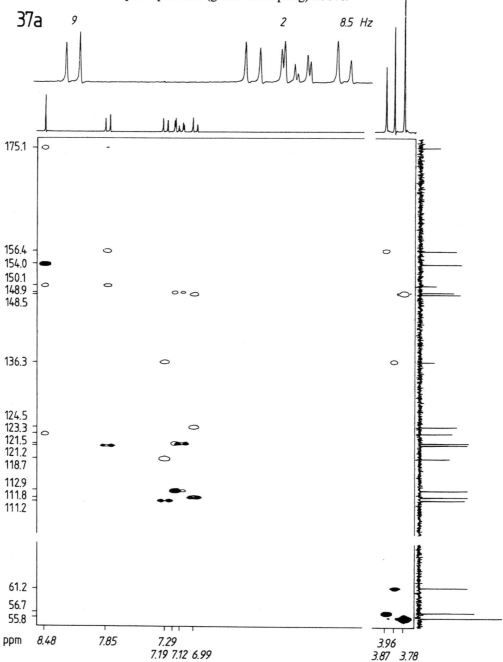

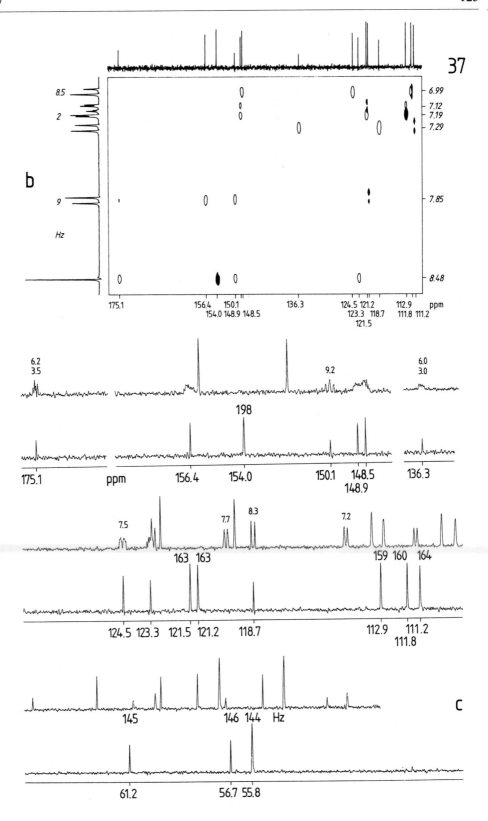

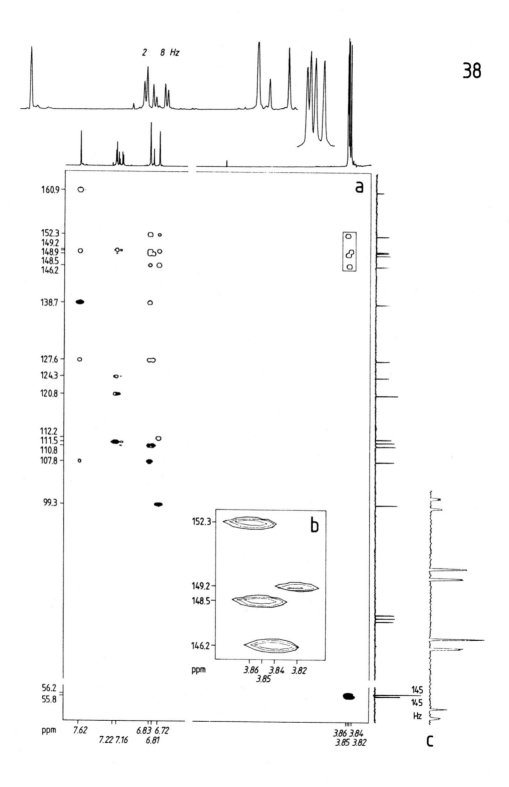

38

38 3′,4′,6,7-Tetramethoxyisoflavone (*3*) was the target of the cyclisation reaction of 3,4-dimethoxyphenol (*1*) with formyl-(3,4-dimethoxyphenyl)acetic acid (*2*) in the presence of polyphosphoric acid.

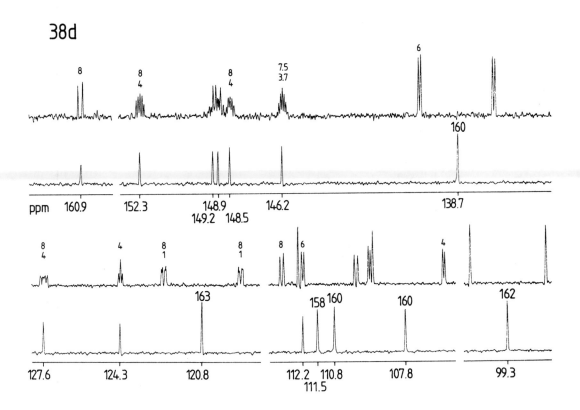

A pale yellow, crystalline product is obtained which fluoresces intense blue and gives the NMR results **38**. Does the product have the desired structure?

Conditions: CDCl$_3$, 25 °C, 200 MHz (*^{1}H*), 50 MHz (*^{13}C*). (a) *CH* COSY (shaded contours) and *CH* COLOC diagrams (unshaded contours) in one diagram with enlarged section (b), and with expanded methoxy quartets (c); (d) sections of ^{13}C NMR spectra, each with *^{1}H* broadband decoupled spectrum below and NOE enhanced coupled spectrum (gated coupling) above.

38d

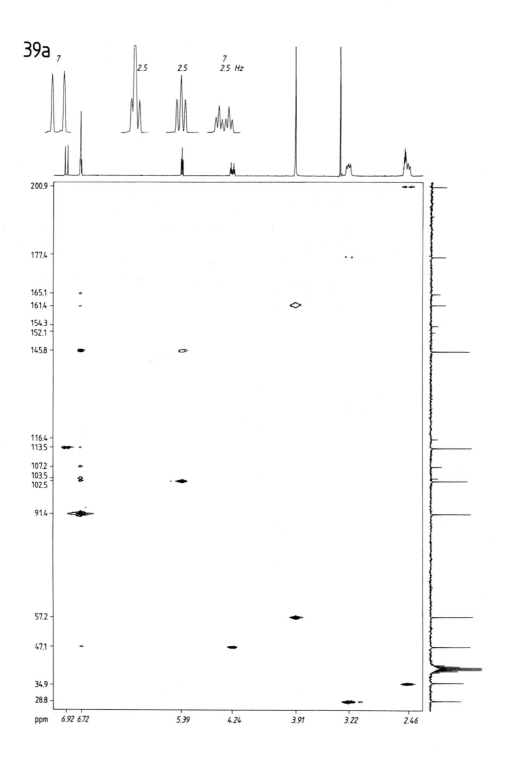

39 An aflatoxin is isolated from *Aspergillus flavus*. Which of the three aflatoxins, B_1, G_1 or M_1, is it given the set of NMR experiments **39**?

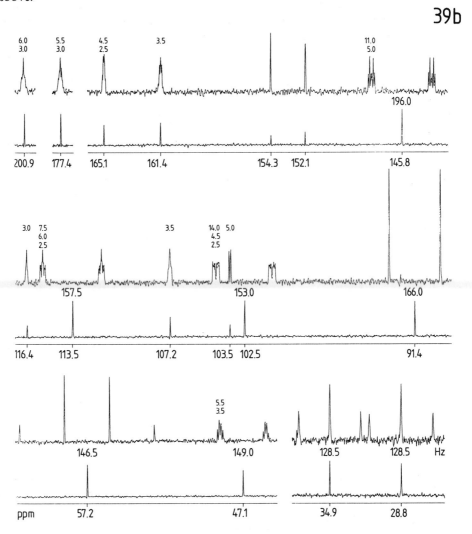

Aflatoxin B_1 Aflatoxin G_1 Aflatoxin M_1

Conditions: $(CD_3)_2SO$, 25 °C, 200 MHz (^1H), 50 MHz (^{13}C). (a) *CH* COSY (shaded contours) and *CH* COLOC diagrams (unshaded contours) in one diagram with expanded ^1H multiplets; (b) sections of ^{13}C NMR spectra, in each case with ^1H broadband decoupled spectrum below and NOE enhanced spectrum (gated decoupling) above.

40 From the plant *Escallonia pulverulenta* (Escalloniaceae), which grows in Chile, an iridoid glucoside of elemental composition $C_{18}H_{22}O_{11}$ was isolated. Formula *1* gives the structure of the iridoid glucoside skeleton.[39]

30 mg of the substance was available and from this the set of NMR results **40** was recorded. What structure does this iridoside have?

Conditions: $(CD_3)_2SO$, 25 °C, 400 and 600 MHz (^1H), 100 MHz (^{13}C). (a) *HH* COSY plot (600 MHz) following D_2O exchange; (b) ^1H NMR spectra before and after deuterium exchange; (c) sections of the ^{13}C NMR spectra, in each case with the ^1H broadband decoupled spectrum below and NOE enhanced decoupled (gated decoupled) spectrum above; (d) *CH* COSY plot with DEPT subspectra for analysis of the *CH* multiplicities; (e) *CH* COLOC diagram.

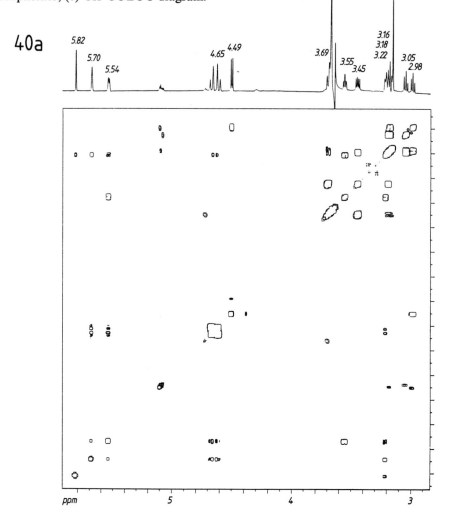

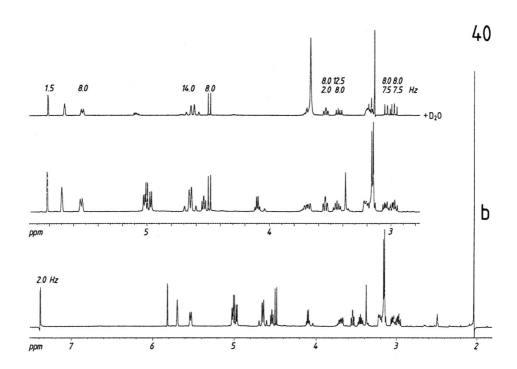

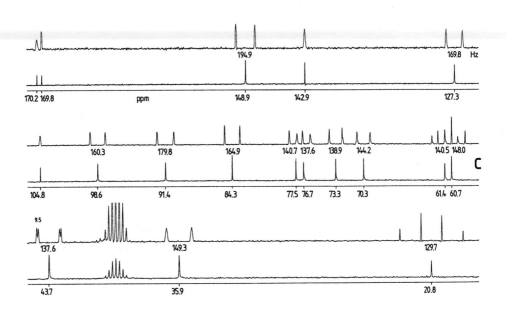

40d

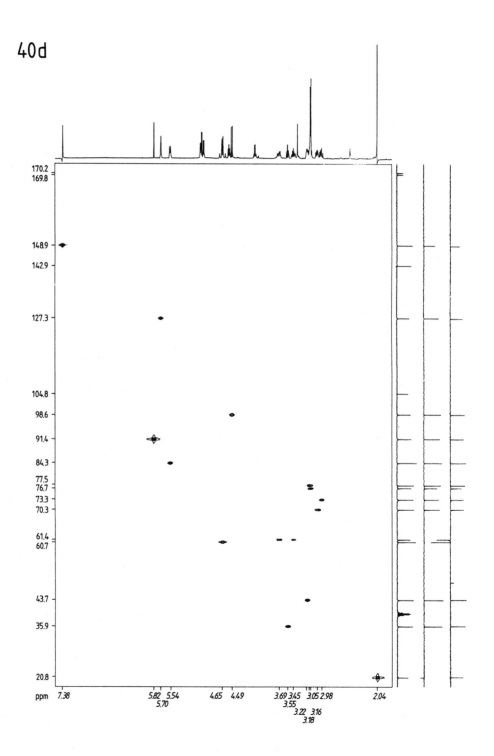

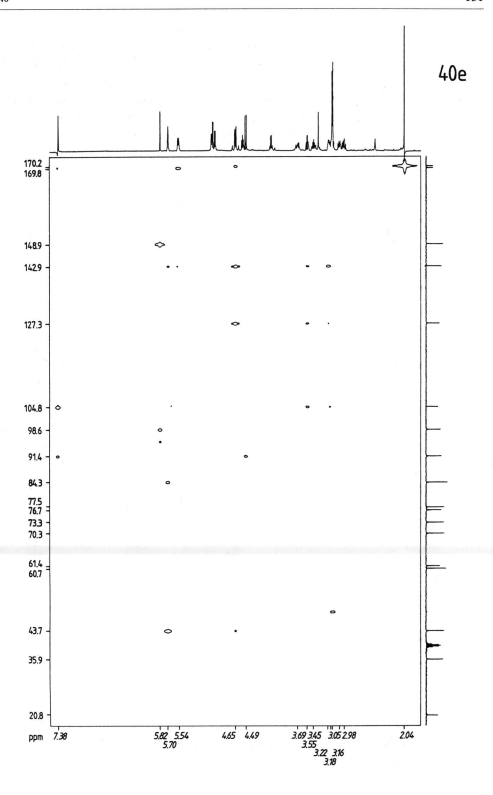

40e

41 A compound with the elemental composition $C_{15}H_{22}O_3$ was isolated from the methanol extract of the Chilean medicinal plant *Centaurea chilensis* (Compositae).[40] What is the structure of the natural product, given the NMR experiments **41**?

Conditions: CD_3OD, 15 mg per 0.3 ml, 400 MHz (^1H), 100 MHz (^{13}C). (a) *HH* COSY plot from *1.2* to *3.5 ppm*; (b) expanded ^1H multiplets from *1.23* to *3.42 ppm*; (c) *CH* COSY diagram from 6 to 130 ppm with ^1H broadband decoupled ^{13}C NMR spectrum (d), DEPT *CH* subspectrum (e) and DEPT spectrum (f) (*CH* and CH_3 positive, CH_2 negative); (g) *CH* COLOC plot.

41a

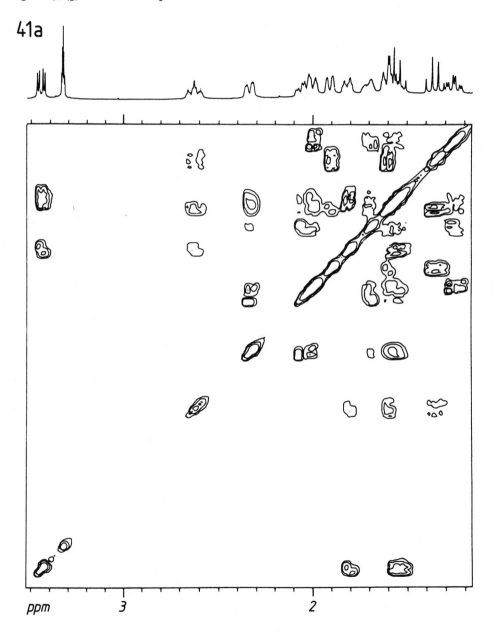

41b

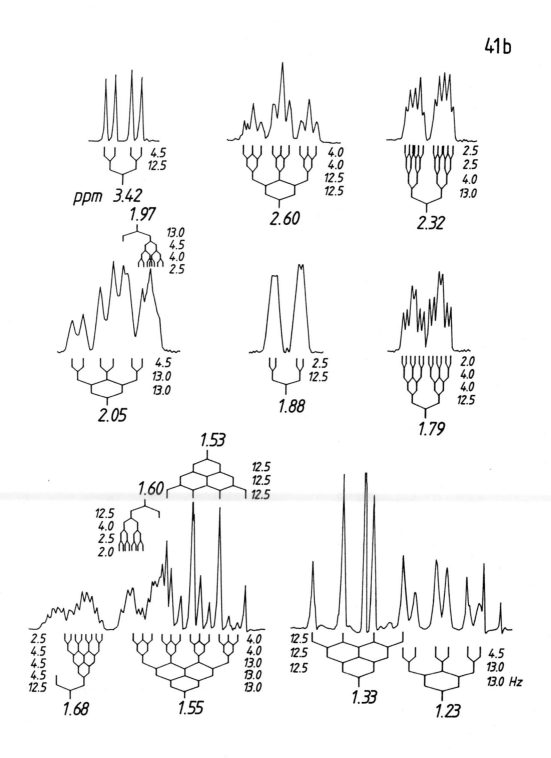

41

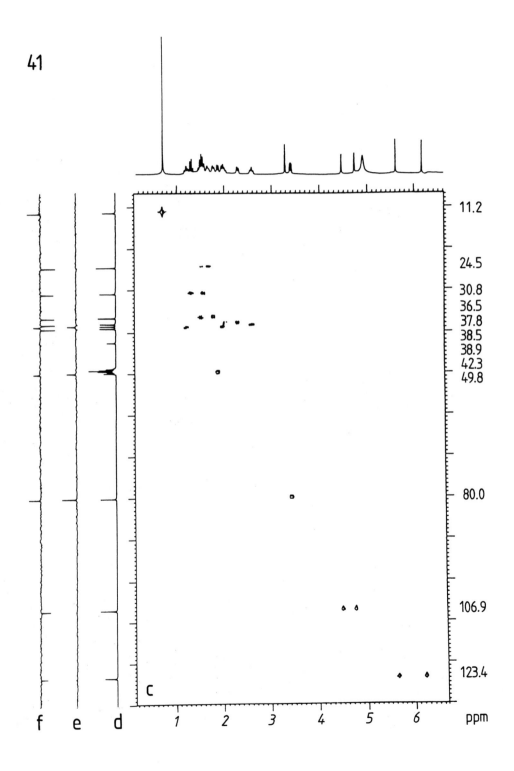

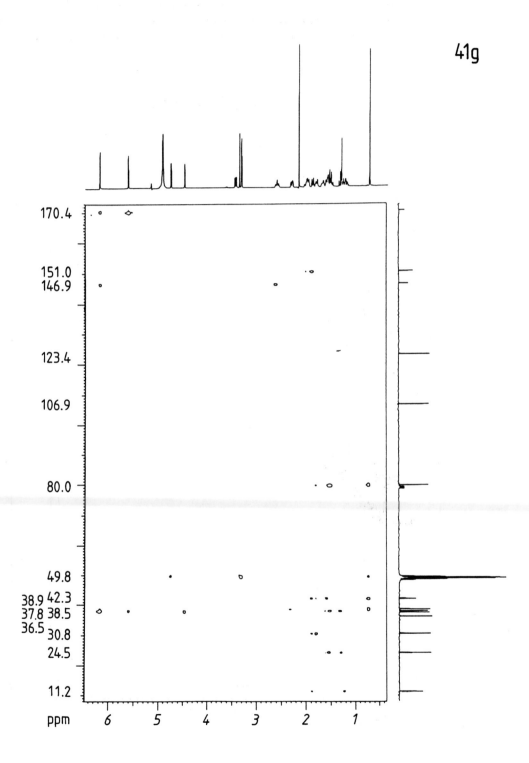

41g

42   The umbelliferone ether structure A was suggested for a natural product which was isolated from galbanum resin.[41] Does this structure fit the NMR results **42**? Is it possible to give a complete spectral assignment despite lack of resolution of the proton signals at 200 MHz? What statements can be made about the relative configuration?

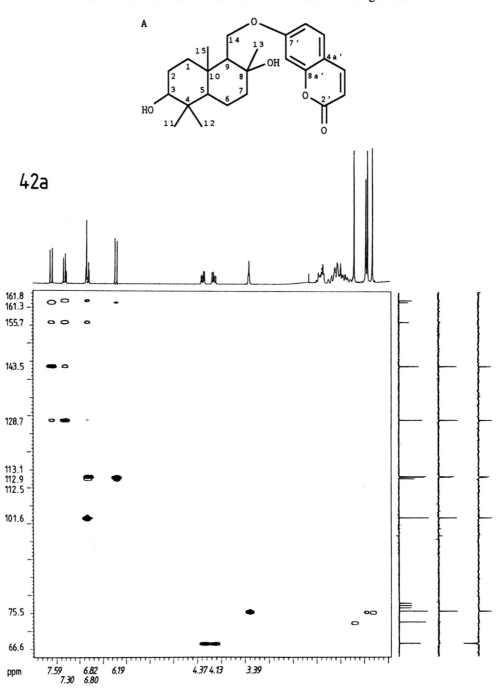

Conditions: CDCl$_3$, 50 mg per 0.3 ml, 25 °C, 200 and 400 MHz (*^{1}H*), 50 and 100 MHz (*^{13}C*). (a, b) C*H* COSY (shaded contours) and C*H* COLOC plots (unshaded contours) in one diagram with DEPT subspectra for identification of the C*H* multiplets; (a) ^{13}C shift range from 66.6 to 161.8 ppm; (b) ^{13}C shift range from 16.0 to 75.5 ppm; (c) sections of ^{13}C NMR spectra (100 MHz), *^{1}H* broadband decoupled spectrum below and NOE enhanced coupled spectrum (gated decoupling) above, with expanded multiplets in the aromatic range; (d) *^{1}H* NMR spectrum with expanded multiplets, integral and NOE difference spectra (irradiated at *0.80, 0.90, 0.96, 1.19, 3.39* and *4.13/4.37 ppm*).

42b

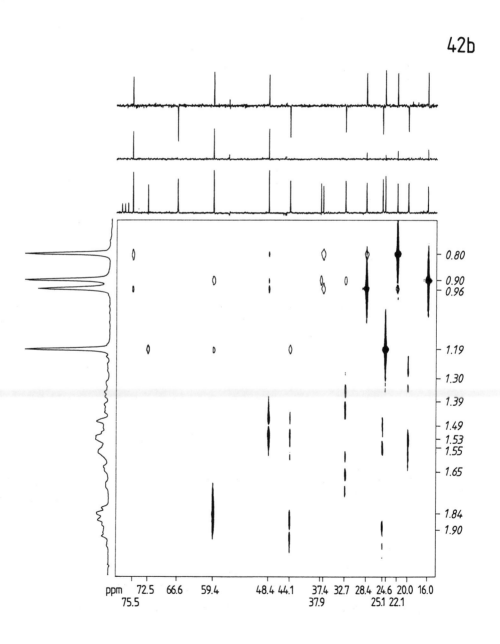

42c

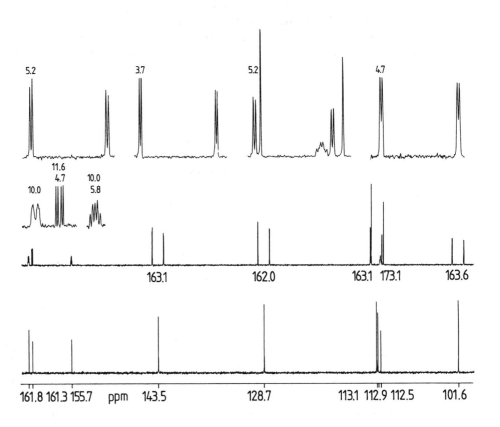

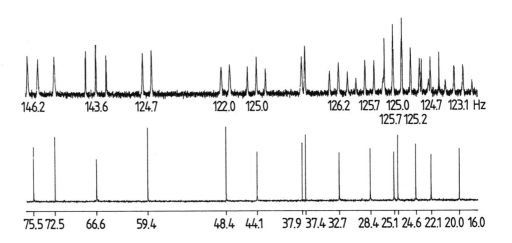

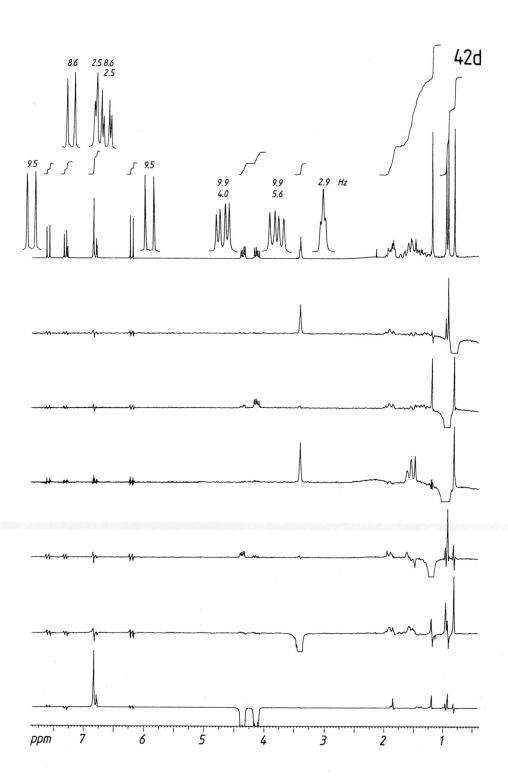

43 A natural product isolated from the plant *Euryops arabicus*, native to Saudi Arabia, has the elemental composition $C_{15}H_{16}O$ determined by mass spectrometry. What is its structure, given the NMR experiments **43**?

Conditions: $CDCl_3$, 20 mg per 0.3 ml, 25 °C, 400 MHz (^1H), 100 MHz (^{13}C). (a) ^1H NMR spectrum with expanded multiplets; (b) sections of ^{13}C NMR spectra, in each case with ^1H broadband decoupled spectrum below and NOE enhanced coupled spectrum (gated decoupling) above; (c) CH COSY and CH COLOC plots in one diagram with DEPT subspectra to facilitate analysis of the CH multiplicities; (d) enlarged section of (c).

43a

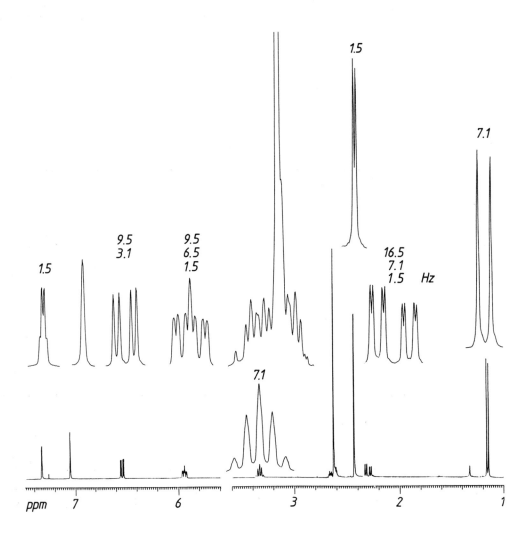

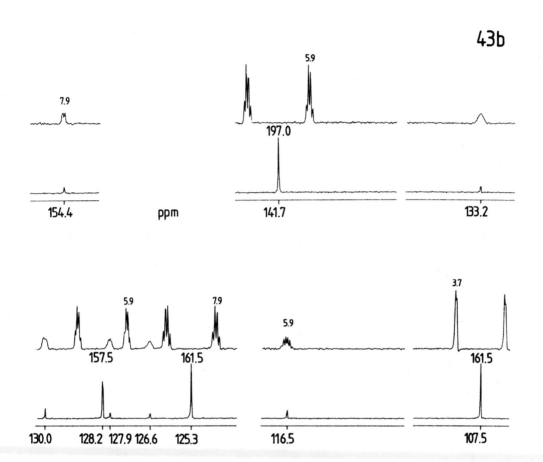

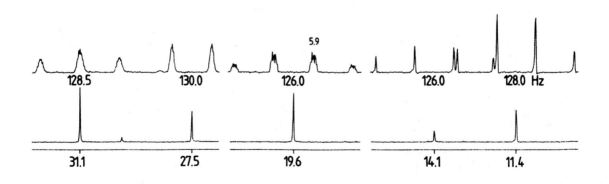

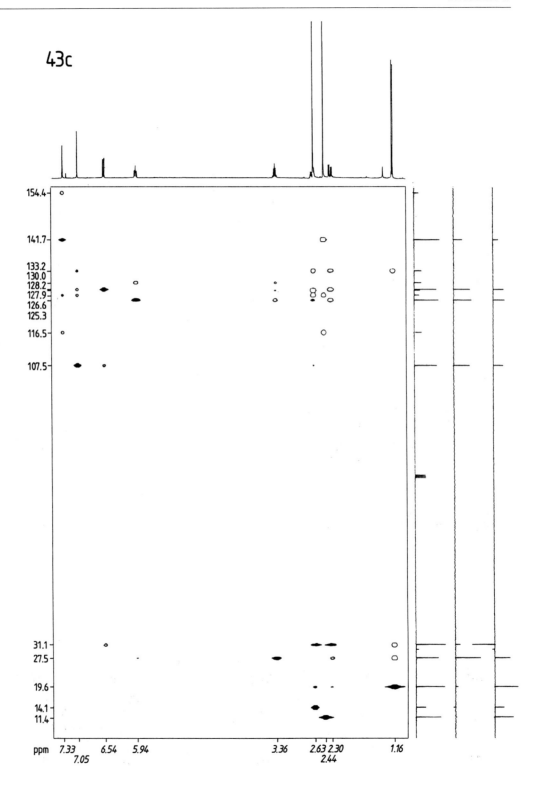

43c

43d

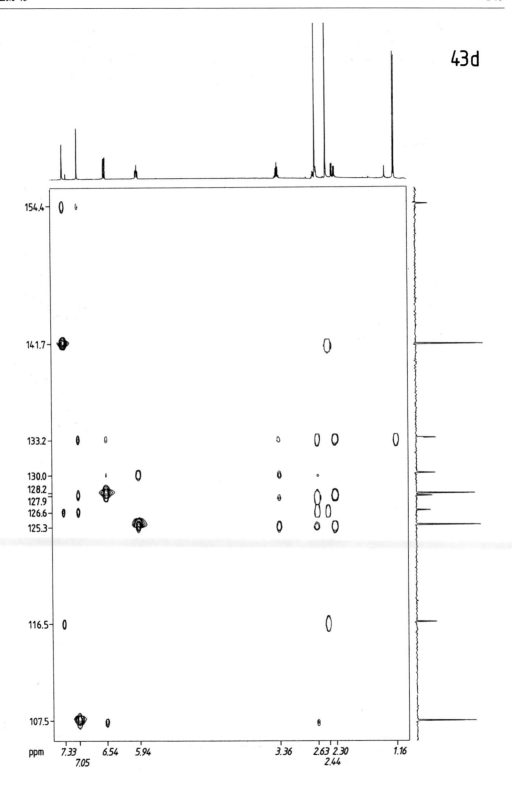

44 A compound with the elemental composition $C_{21}H_{28}O_6$, determined by mass spectrometry, was isolated from the light petroleum extract of the leaves of *Senecio darwinii* (Compositae, Hooker and Arnolt), a plant which grows in Tierra del Fuego. What structure can be derived from the set of NMR experiments **44**?

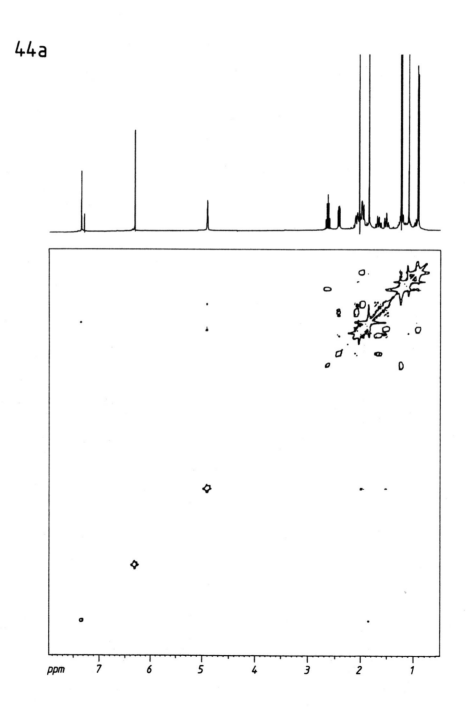

44a

Conditions: CDCl$_3$, 25 mg per 0.3 ml, 25 °C, 400 MHz (^1H), 100 MHz (^{13}C). (a) *HH* COSY plot; (b) *HH* COSY plot, section from *0.92* to *2.62 ppm*; (c) *CH* COSY plot with DEPT subspectra to facilitate analysis of the *CH* multiplets; (d) *CH* COSY plot, sections from *0.92* to *2.62* and *8.8* to *54.9 ppm*; (e) *CH* COLOC plot; (f) ^1H NMR spectrum with expanded multiplets and NOE difference spectra, irradiations at *1.49*, *1.66*, *2.41* and *6.29 ppm*.

44b

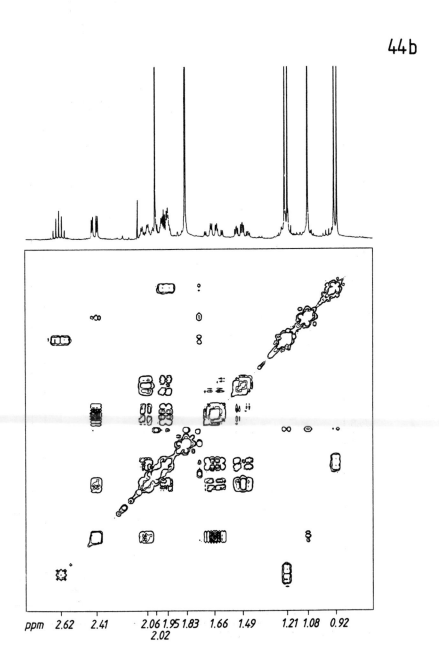

ppm   2.62   2.41   2.06 1.95 1.83   1.66   1.49   1.21 1.08   0.92
                            2.02

44c

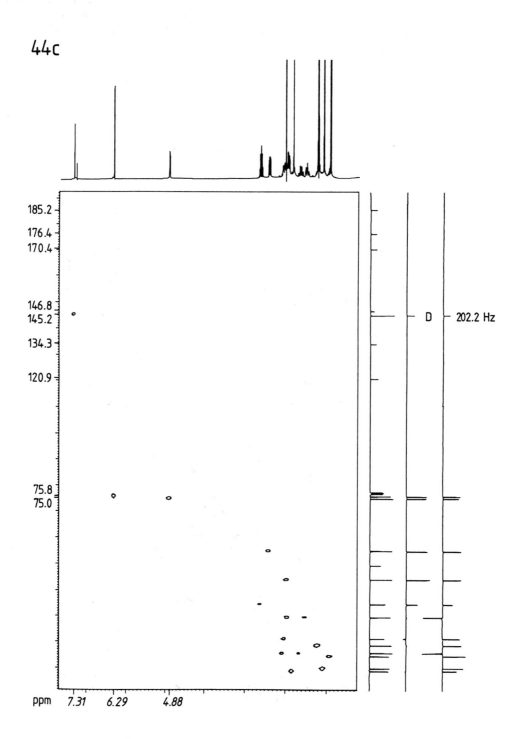

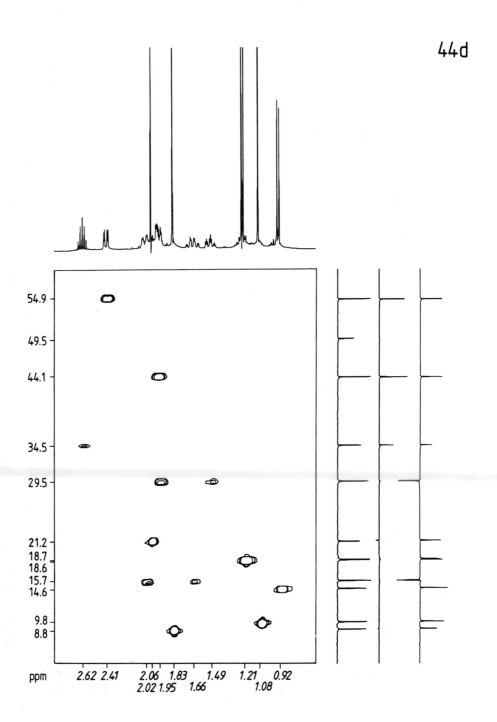

44d

44e

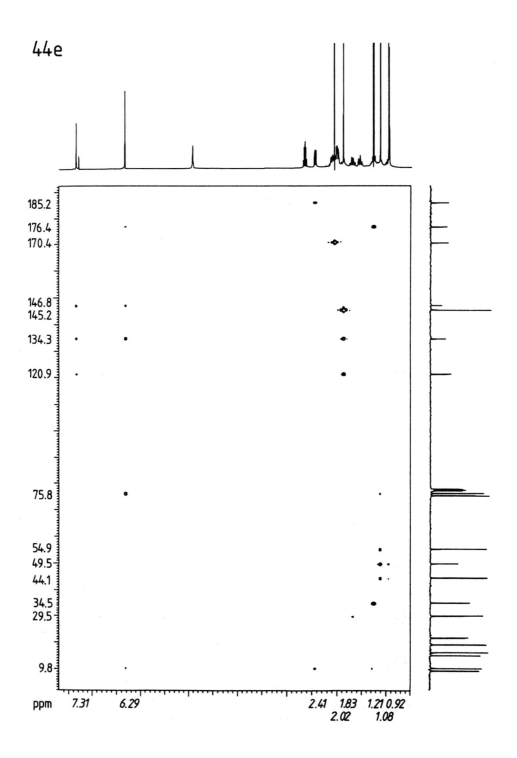

44f

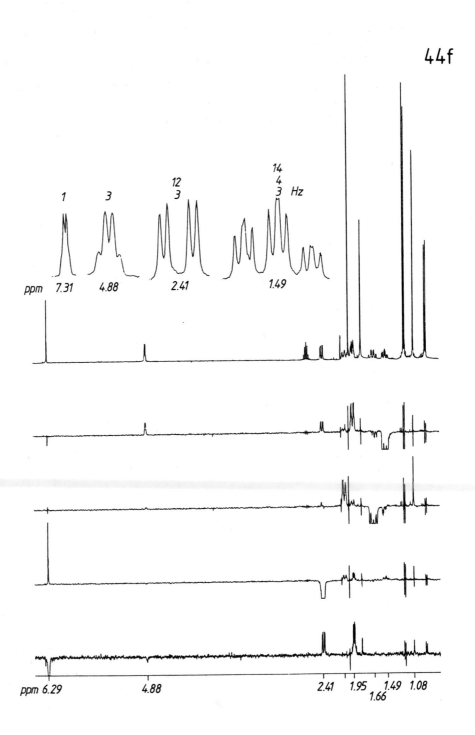

45 A substance with the molecular formula $C_{17}H_{20}O_4$, as determined by mass spectrometry, was isolated from the light petroleum extract of the Chilean medicinal plant *Centaurea chilensis* (Compositae); 8 mg were available for the set of NMR experiments **45**. Beyond the shift range shown in (c), the ^{13}C NMR spectrum shows the signals of quaternary C atoms at 170.1, 169.2, 149.8, 142.9 and 137.5 ppm. A CH COLOC plot was not recorded owing to shortage of material and time. It was nonetheless possible to identify the natural product which was already known.[43] What is its structure?

Conditions: $CDCl_3$, 8 mg per 0.3 ml, 25 °C, 400 MHz (^{13}C). (a) Expanded ^1H multiplets; (b) *HH* COSY plot; (c) *CH* COSY plot with a DEPT subspectrum to distinguish CH_2 (negative) from CH and CH_3 (both positive).

45a

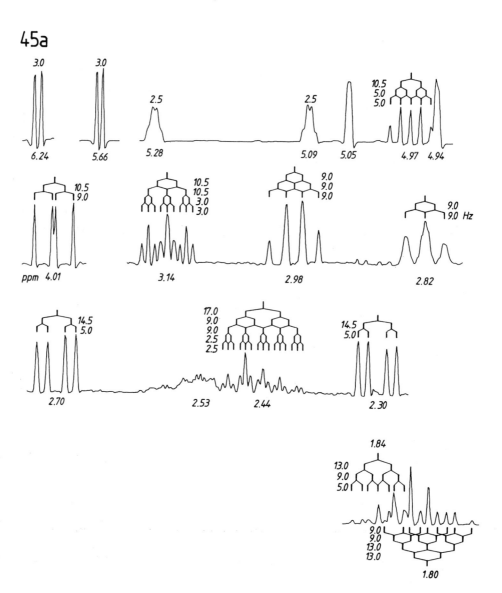

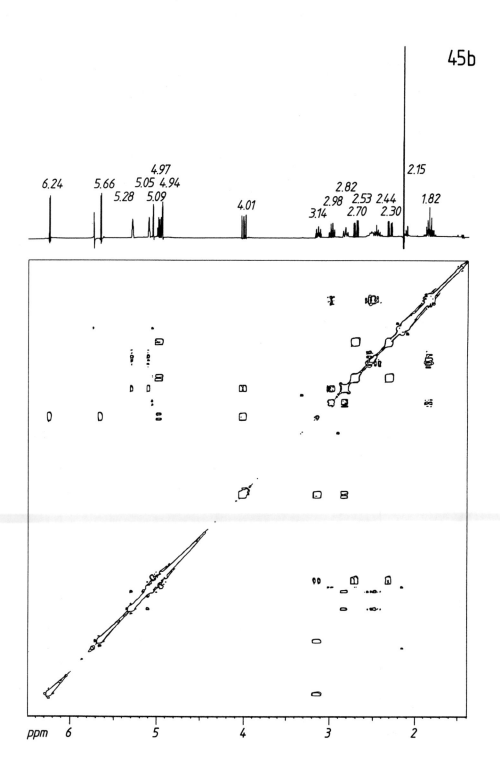

45b

45c

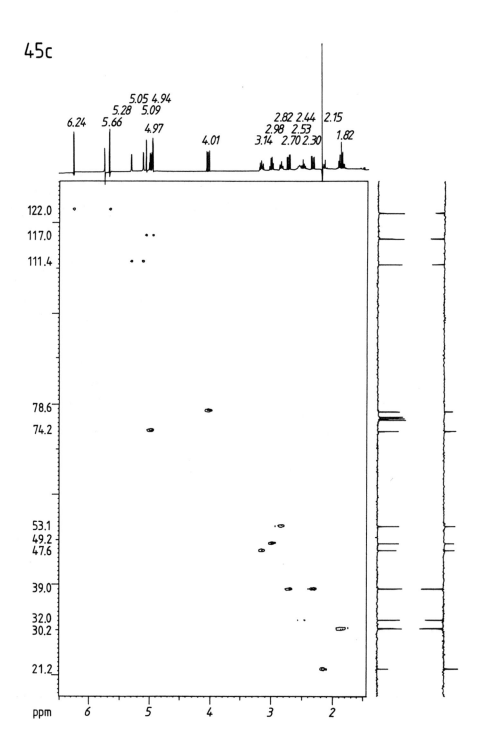

46 Sapogenins of the dammaran type were isolated from the leaves and roots of the plant *Panax notoginseng*, native to China.[44] One of these sapogenins has the elemental composition $C_{30}H_{52}O_4$ and produces the set of NMR results **46**. What is the structure of the sapogenin?

Conditions: 20 mg, $CDCl_3$, 20 mg per 0.3 ml, 25 °C, 200 and 400 MHz (^1H), 100 MHz (^{13}C). (a) *HH* COSY plot (400 MHz) with expansion of multiplets; (b) NOE ^1H difference spectra (200 MHz), decoupling of the methyl protons shown; (c) *CH* COSY plot with enlarged section (d); (e) *CH* COLOC plot.

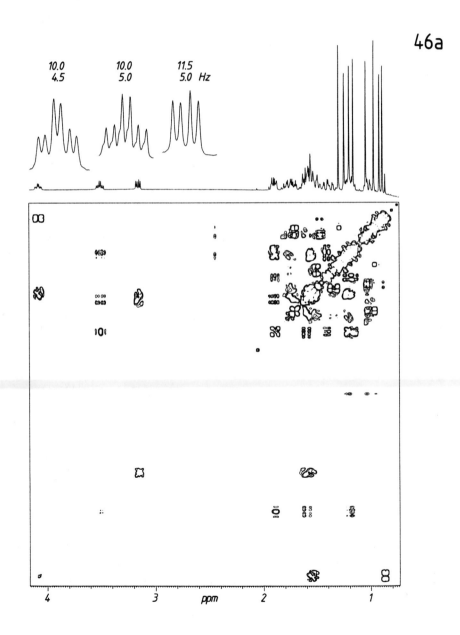

46a

46b

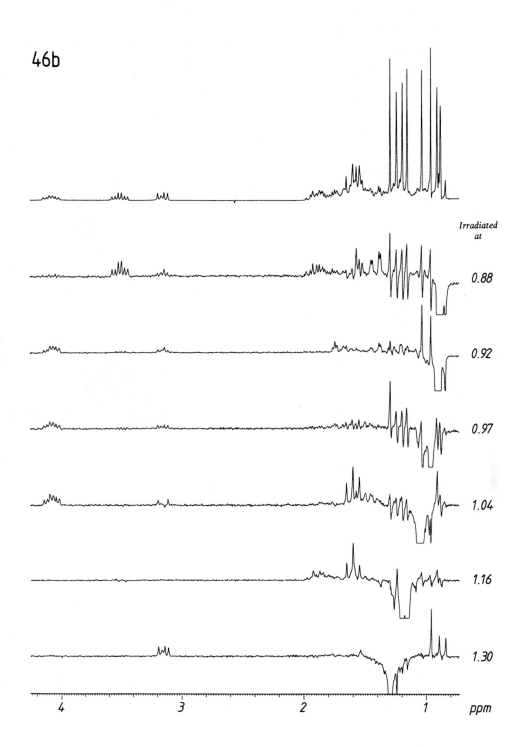

Irradiated at

0.88

0.92

0.97

1.04

1.16

1.30

46c

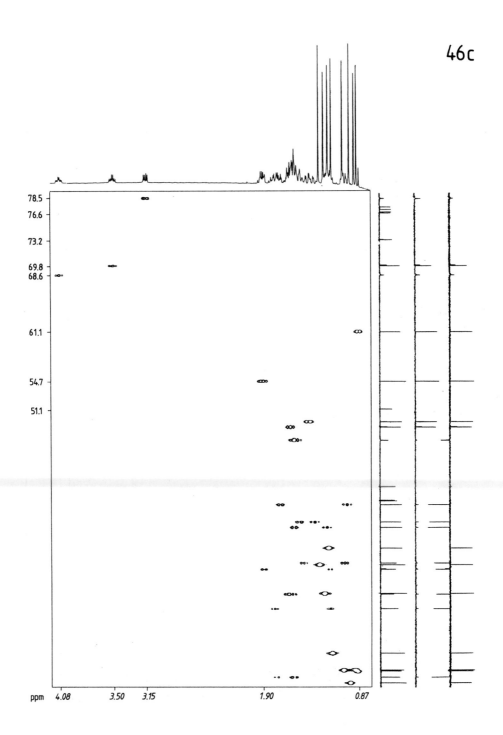

46d

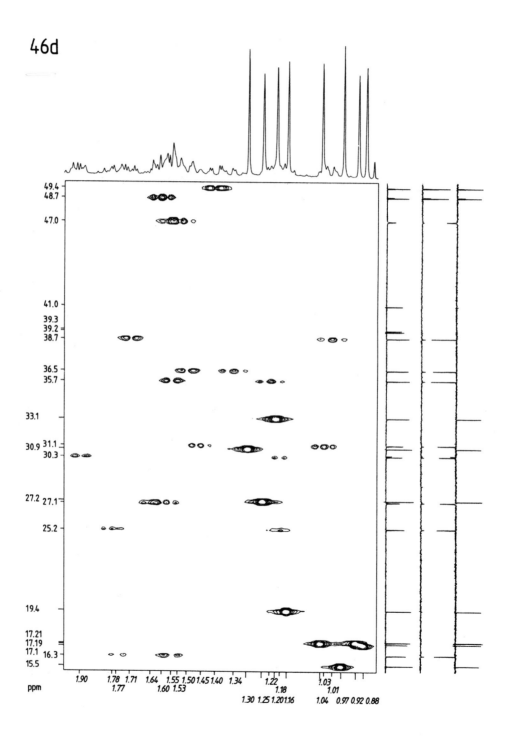

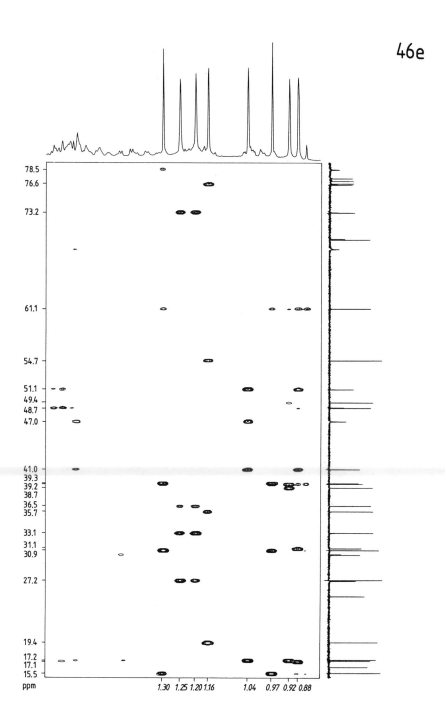

46e

47 An 8 mg amount of an alkaloid was isolated from the plant *Picrasma quassioides* Bennet (Simaroubaceae),[45] native to East Asia. The formula $C_{16}H_{12}N_2O_3$ was obtained by mass spectrometry. What is the structure of the alkaloid given the NMR results **47**?

Conditions: CDCl$_3$, 8 mg per 0.3 ml, 25 °C, 200 MHz (*^{1}H*), 50 and 100 MHz, respectively (*^{13}C*). (a) *^{1}H* NMR spectrum with expanded partial spectrum (*7.48–8.74 ppm*) and *HH* COSY plot inset; (b) *^{13}C* NMR partial spectra, each with *^{1}H* decoupled spectrum below and NOE enhanced coupled spectrum (gated decoupling) above and with the expanded multiplets from 116.0 to 133.1 and 138.4 to 157.7 ppm; (c) *CH* COSY and *CH* COLOC plots in one diagram, with *CH* COSY correlation signals encircled. Correlation signals in the aliphatic range (61.5/*4.34* and 60.7/*3.94* ppm) are not shown for reasons of space.

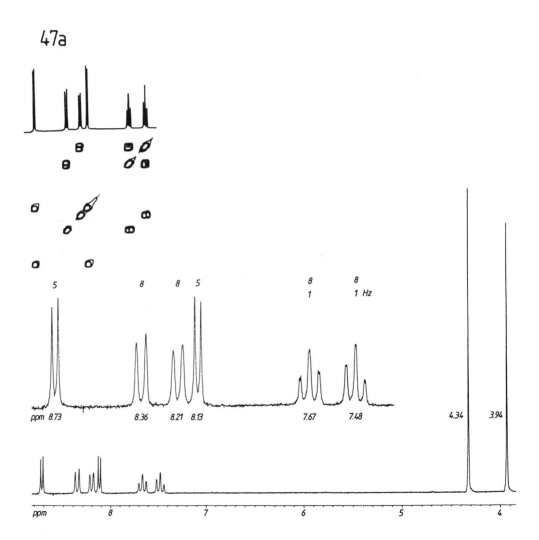

47a

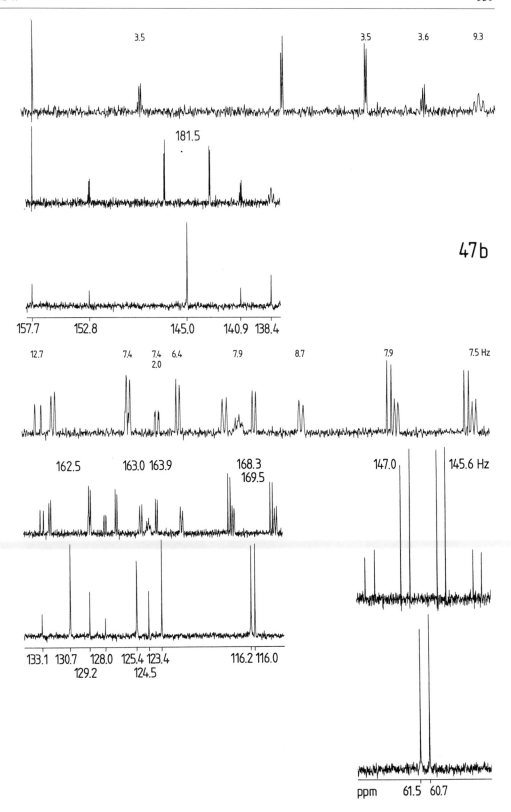

47 b

47c

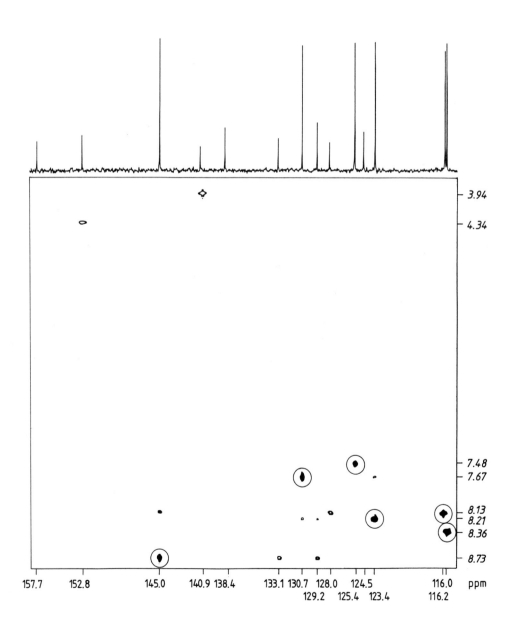

48 The hydrochloride of a natural product which is intoxicating and addictive produced the set of NMR results **48**. What is the structure of the material? What additional information can be derived from the NOE difference spectrum?

Conditions: CD$_3$OD, 30 mg per 0.3 ml, 25 °C, 400 MHz (^1H), 100 MHz (^{13}C). (a) *HH* COSY plot; (b) ^1H NMR spectrum with NOE difference spectrum, irradiation at *2.92 ppm*; (c) ^{13}C NMR spectra, each with the ^1H broadband decoupled spectrum below and NOE enhanced coupled spectrum (gated decoupling) above; (d) *CH* COSY and *CH* COLOC plots in one diagram with enlarged section (*64.5–65.3/2.44–3.56 ppm*).

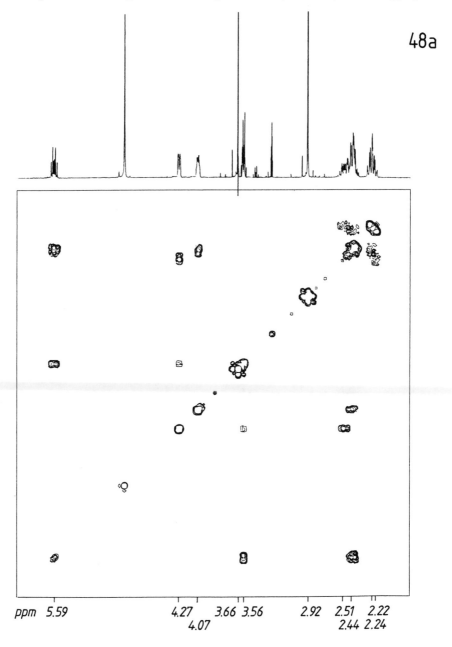

48a

ppm 5.59   4.27   3.66 3.56   2.92   2.51 2.22
                          4.07   2.44 2.24

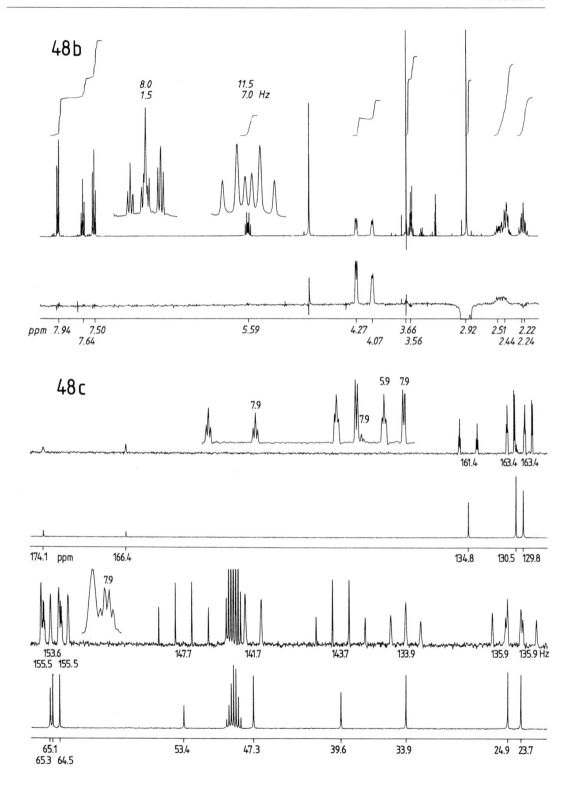

48b

8.0
1.5

11.5
7.0 Hz

ppm 7.94   7.50
          7.64

5.59

4.27
      4.07

3.66
3.56

2.92

2.51
2.44

2.22
2.24

48c

7.9

7.9

5.9 7.9

161.4   163.4 163.4

174.1 ppm   166.4

134.8   130.5 129.8

7.9

153.6
155.5 155.5

147.7

141.7

143.7

133.9

135.9 135.9 Hz

65.1
65.3 64.5

53.4

47.3

39.6

33.9

24.9 23.7

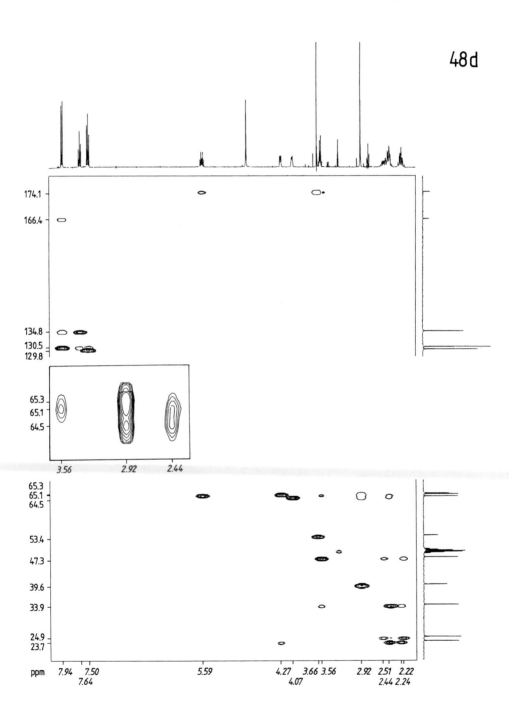

48d

49 Amongst products isolated from *Heliotropium spathulatum* (Boraginaceae) were 9 mg of a new alkaloid which gave a positive Ehrlich reaction with *p*-dimethylamino-benzaldehyde. The molecular formula determined by mass spectrometry is $C_{15}H_{25}NO_5$. What is the structure of the alkaloid given the set of NMR results **49**? Reference 31 is useful in providing the solution to this problem.

Conditions: CDCl$_3$, 9 mg per 0.3 ml, 25 °C, 400 MHz (^1H), 100 MHz (^{13}C). (a) *HH* COSY plot; (b) ^1H NMR partial spectrum beginning at *1.98 ppm* with NOE difference spectra, irradiations at the given chemical shifts; (c) *CH* COSY plot with DEPT spectrum (*CH* and *CH$_3$* positive, *CH$_2$* negative); (d) *CH* COLOC plot.

49a

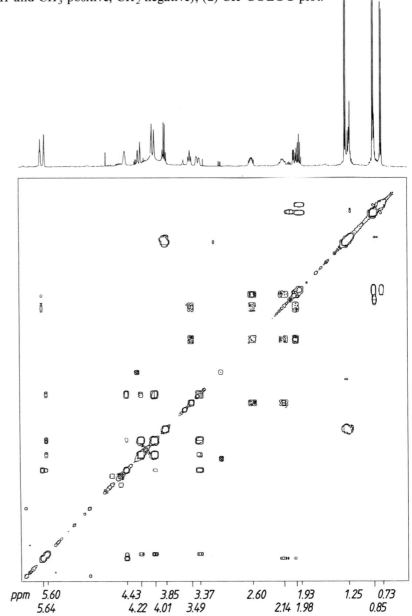

| ppm | 5.60 | | 4.43 | 3.85 | 3.37 | 2.60 | 1.93 | 1.25 | 0.73 |
| | 5.64 | | 4.22 | 4.01 | 3.49 | | 2.14 1.98 | | 0.85 |

49b

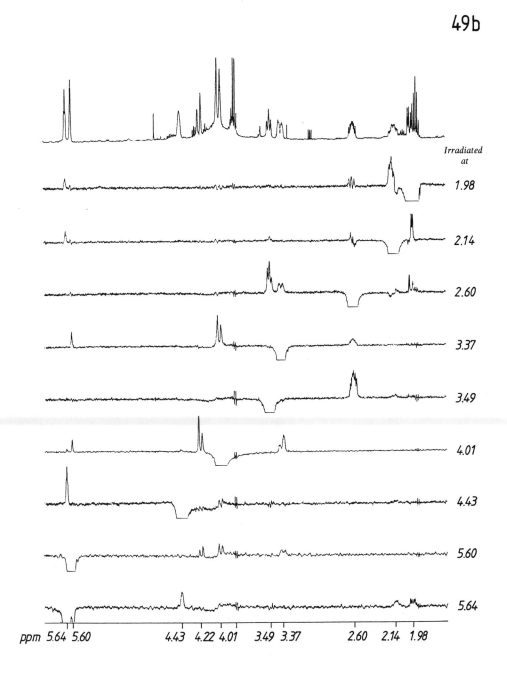

Irradiated at

1.98

2.14

2.60

3.37

3.49

4.01

4.43

5.60

5.64

ppm 5.64 5.60   4.43 4.22 4.01   3.49 3.37   2.60   2.14 1.98

49c

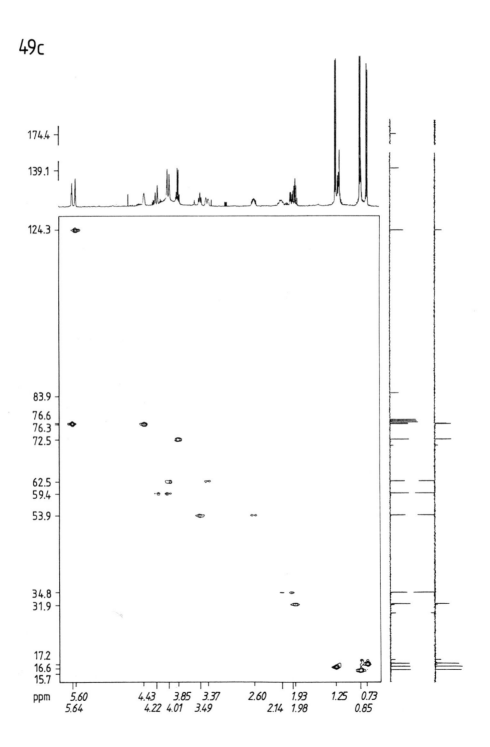

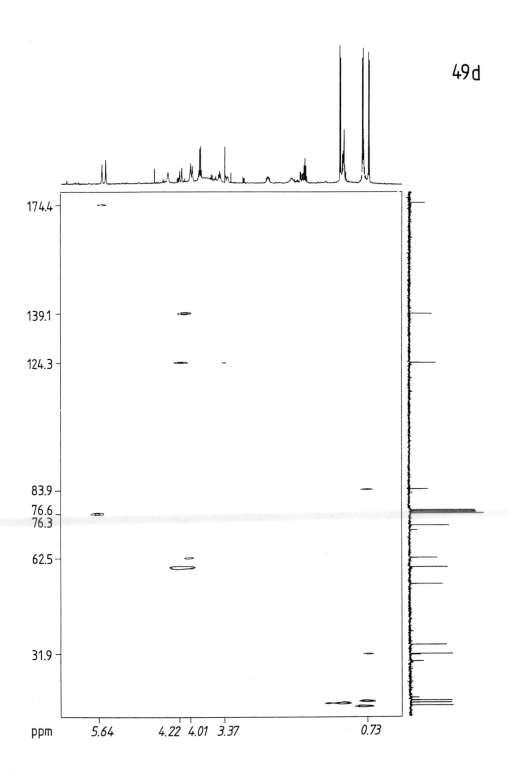

49d

50   A compound with the molecular formula $C_7H_{13}NO_3$, determined by mass spectrometry, was isolated from the plant *Petiveria alliacea* (Phytolaccaceae). What is its structure given the set of NMR results **50**?

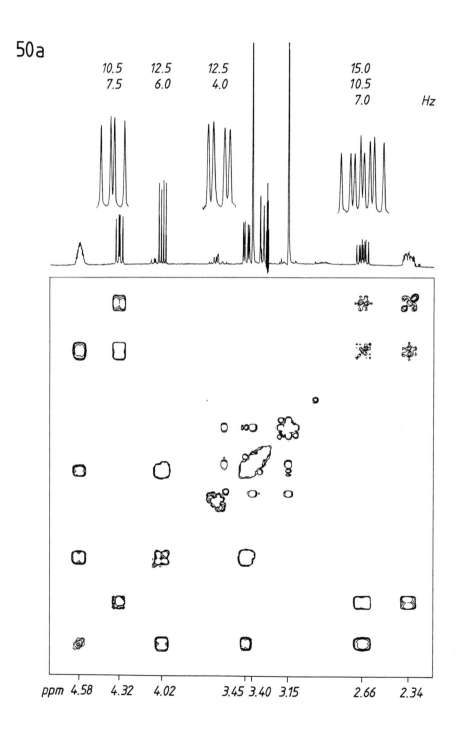

Conditions: CD$_3$OD, 30 mg per 0.3 ml, 25 °C, 400 and 200 MHz (^1H), 100 MHz (^{13}C). (a) *HH* COSY plot with expanded ^1H multiplets; (b) ^1H NOE difference spectra, decoupling at the signals indicated, 200 MHz; (c) *CH* COSY and *CH* COLOC plots in one diagram with DEPT spectrum (*CH$_2$* negative, *CH* and *CH$_3$* positive) and coupled (gated decoupling) ^{13}C NMR spectrum above.

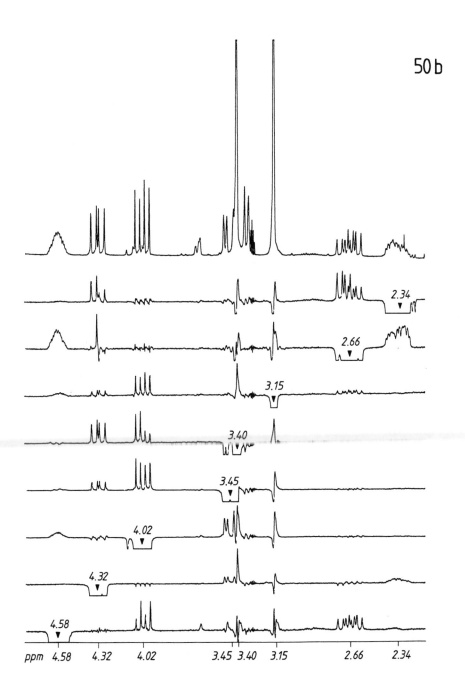

50 b

50c

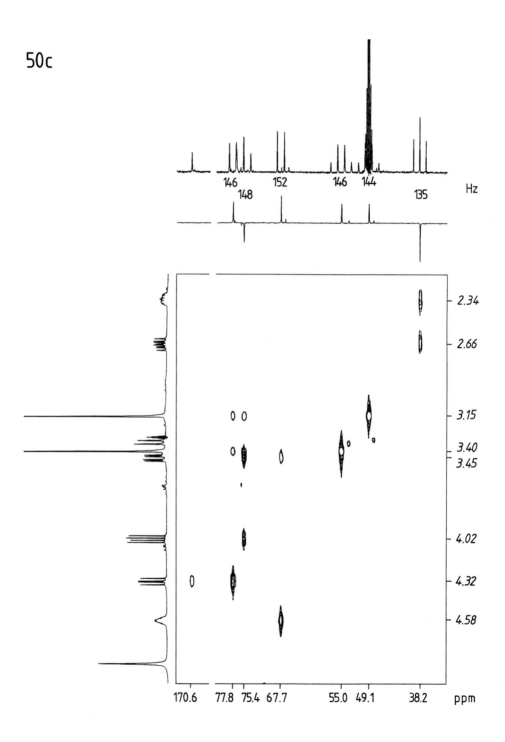

4 SOLUTIONS TO PROBLEMS 1–50

1 Dimethyl *cis*-cyclopropane-1,2-dicarboxylate

In the ^1H NMR spectrum it is possible to discern a four-spin system of the AMX_2 type for the three different kinds of proton of the *cis*-1,2-disubstituted cyclopropane ring. The X_2 protons form a doublet of doublets with *cis* coupling $^3J_{AX}$ (8.5 Hz) and *trans* coupling $^3J_{MX}$ (6.7 Hz). The signals of protons H^A and H^M also show *geminal* coupling $^2J_{AM}$ (5.1 Hz) and splitting into triplets (two H^X) of doublets. The *trans* isomer would show a four-spin system of the type $AA'BB'$ or $AA'XX'$ (according to the shift difference relative to the coupling constant).

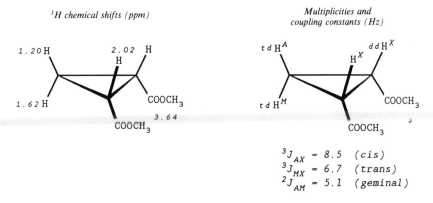

^1H chemical shifts (ppm)

1.20 H
2.02 H
1.62 H
COOCH$_3$
3.64
COOCH$_3$

Multiplicities and coupling constants (Hz)

$td\ H^A$
$dd\ H^X$
H^X
COOCH$_3$
$td\ H^M$
COOCH$_3$

$$^3J_{AX} = 8.5 \quad (cis)$$
$$^3J_{MX} = 6.7 \quad (trans)$$
$$^2J_{AM} = 5.1 \quad (geminal)$$

2 Ethyl acrylate, $C_5H_8O_2$

The empirical formula implies two double-bond equivalents. The ^1H NMR spectrum shows multiplet systems whose integral levels are consistent with the eight H atoms of the empirical formula in a ratio of 3:2:3. The triplet at *1.3* and the quartet at *4.2 ppm* with the common coupling constant of *7 Hz* belong to the A_3X_2 system of an ethoxy group, $-OCH_2CH_3$. Three alkene protons between *5.7* and *6.6 ppm* with the *trans*, *cis* and *geminal* couplings (*13, 8* and *2 Hz*) which are repeated in their coupling partners, identify the *ABC* system of a vinyl group, $-CH=CH_2$. If one combines both structural elements $(C_2H_5O + C_2H_3 = C_4H_8O)$ and compares the result with the empirical formula $(C_5H_8O_2)$, then C and O as missing atoms give a CO double bond in accordance with the second double-bond equivalent. Linking the structural elements together leads to ethyl acrylate.

Note how dramatically the roofing effects of the *AB* and *BC* part systems change the intensities of the doublet of doublets of proton *B* in spectrum **2**.

3   *cis*-1-Methoxybut-1-en-3-yne, C_5H_6O

Three double-bond equivalents which follow from the empirical formula can be confirmed in the ^1H NMR spectrum using typical shifts and coupling constants. The ^1H signal at 3.05 ppm indicates an ethynyl group, $-C\equiv C-H$ (H^A); an *MX* system in the alkene shift range with $\delta_M = 4.50$ and $\delta_X = 6.30\,ppm$, respectively, and the coupling constants $^3J_{MX} = 8\,Hz$, reveal an ethene unit $(-CH=CH-)$ with a *cis* configuration of the protons. The intense singlet at *3.8 ppm* belongs to a methoxy group, $-OCH_3$, whose $-I$ effect deshields the H^X proton, whilst its $+M$ effect shields the H^M proton. The bonding between the ethenyl and ethynyl groups is reflected in the long-range couplings $^4J_{AM} = 3$ and $^5J_{AX} = 1\,Hz$.

4   *trans*-3-(*N*-Methylpyrrol-2-yl)propenal, C_8H_9NO

The empirical formula contains five double-bond equivalents. In the ^1H NMR spectrum a doublet signal at *9.55 ppm* stands out. This chemical shift value would fit an aldehyde function. Since the only oxygen atom in the empirical formula is thus assigned a place, the methyl signal at *3.80 ppm* does not belong to a methoxy group, but rather to an *N*-methyl group.

The coupling constant of the aldehyde doublet (*7.8 Hz*) is repeated in the doublet of doublets signal at *6.3 ppm*. Its larger splitting of *15.6 Hz* is observed also in the doublet at *7.3 ppm* and indicates a CC double bond with a *trans* configuration of the *vicinal* protons.

The coupling of 7.8 Hz in the signals at *9.55* and *6.3 ppm* identifies the (E)-propenal part structure. The ^1H shift thus reflects the $-M$ effect of the conjugated carbonyl group.

Apart from the N-methyl group, three double-bond equivalents and three multiplets remain in the chemical shift range appropriate for electron rich heteroaromatics, *6.2* to *6.9 ppm*. N-Methylpyrrole is such a compound. Since in the multiplets at *6.25* and *6.80 ppm* the $^3J_{HH}$ coupling of *4.0 Hz* is appropriate for pyrrole protons in the 3- and 4-positions, the pyrrole ring is deduced to be substituted in the 2-position.

^1H chemical shifts (ppm)   *Multiplicities and coupling constants (Hz)*

$$^3J_{1,2} = 7.8$$
$$^3J_{2,3} = 15.6$$
$$^5J_{3,4'} = 0.5$$
$$^3J_{3',4'} = 4.0$$
$$^4J_{3',4'} = 1.6$$
$$^3J_{4',5'} = 2.5$$

5 1,9-Bis(pyrrol-2-yl)pyrromethane

The multiplets are sorted according to the principle that identical coupling constants identify coupling partners and after comparing *J*-values with characteristic pyrrole $^3J_{HH}$ couplings, two structural fragments **A** and **B** are deduced. The *AB* system ($^3J_{AB} = 4.4$ *Hz*) of doublets (*2.2* and *2.6 Hz*, respectively with NH) at *6.73* (H^A) and *7.00 ppm* (H^B) belongs to the 2,5-disubstituted pyrrole ring **A**. The remaining three multiplets at *6.35*, *6.89* and *7.17 ppm* form an *ABC* system, in which each of the *vicinal* couplings of the pyrrole ring ($^3J_{AB} = 3.7$ and $^3J_{AC} = 2.5$ *Hz*) characterises a 2-monosubstituted pyrrole ring **B**.

^1H chemical shifts (ppm) and multiplicities

11.60 (1x)H^X

dd 6.73 H^A H^{B}7.00 dd

Coupling constants (Hz)
$^3J_{AB} = 4.4$
$^4J_{AX} = 2.2$
$^4J_{BX} = 2.6$

H^{X}12.40 (2x)

—H^{C}7.17 "t"d

ddd 6.89 H^B H^{A}6.35 d"t"

$^3J_{AB} = 3.7$
$^3J_{AC} = 2.5$
$^3J_{CX} = 2.5$
$^4J_{AX} = 2.5$
$^4J_{BX} = 2.5$

Each of the two pyrrole rings occurs twice in the molecule judging by their integrated intensities relative to the methine singlet at *6.80 ppm*. Their connection with the methine group, which itself only occurs once (*6.80 ppm*), gives 1,9-bis(pyrrol-2-yl)pyrromethane **C**, a result which is illuminating in view of the reaction which has been carried out.

6 3-Acetylpyridine, C_7H_7NO

The *^1H* NMR spectrum contains five signals with integral levels in the ratios 1:1:1:1:3; four lie in the shift range appropriate for aromatics or heteroaromatics and the fifth is evidently a methyl group. The large shift values (up to *9.18 ppm*, aromatics) and typical coupling constants (*8* and *5 Hz*) indicate a pyridine ring, which accounts for four out of the total five double-bond equivalents.

Four multiplets between *7.46* and *9.18 ppm* indicate monosubstitution of the pyridine ring, either in the 2- or 3-position but not in the 4-position, since for a 4-substituted pyridine ring an *AA'XX'* system would occur. The position of the substituents follows from the coupling constants of the threefold doublet at *7.46 ppm*, whose shift is appropriate for a β-proton on the pyridine ring (**A**). The *8 Hz* coupling indicates a proton in the γ-position (**B**); the *5 Hz* coupling locates a *vicinal* proton in position α (**C**), the additional *0.9 Hz* coupling locates the remaining proton in position α' (**D**) and thereby the β-position of the substituent.

	5.0Hz H. N	H. N H
N H 7.46ppm	H H 8.0Hz	H H	H H
A	B	C	D

This example shows how it is possible to pin-point the position of a substituent from the coupling constant of a clearly structured multiplet, whose shift can be established beyond doubt. The coupling constants are repeated in the multiplets of the coupling partners; from there the assignment of the remaining signals follows without difficulty.

A monosubstituted pyridine ring and a methyl group add up to C_6H_7N. The atoms C and O which are missing from the empirical formula and a double-bond equivalent indicate a carbonyl group. The only structure compatible with the presence of these fragments is 3-acetylpyridine.

^1H *chemical shifts (ppm)*   *Multiplicities and coupling constants (Hz)*

$$^3J_{4,5} = 8.0$$
$$^3J_{5,6} = 5.0$$
$$^4J_{2,4} = 1.8$$
$$^4J_{4,6} = 2.2$$
$$^5J_{2,5} = 0.9$$

7   6,4'-Dimethoxyisoflavone

Two intense signals at *3.70* and *3.80 ppm* identify two methoxy groups as substituents. The aromatic resonances arise from two subspectra, an *AA'XX'* system and an *ABM* system. The *AA'XX'* part spectrum indicates a *para*-disubstituted benzene ring, and locates a methoxy group in the 4'-position of the phenyl ring *B*. A doublet of doublets with *ortho* and *meta* coupling (*9* and *3 Hz*, respectively) belongs to the *ABM* system, from which a 1,2,4-trisubstituted benzene ring (ring *A*) is derived.

Ring B   Ring A

Hence the solution is isomer 6,4'- or 7,4'-dimethoxyisoflavone **A** or **B**.

The decisive clue is given by the large shift ($7.51\ ppm$) of the proton marked with an asterisk, which only shows one *meta* coupling. This shift value fits structure **A**, in which the $-M$ effect and the anisotropy effect of the carbonyl group lead to deshielding of the proton in question. In **B** the $+M$ effect of the two *ortho* oxygen atoms would lead to considerable shielding. The methoxy resonances cannot be assigned conclusively to specific methoxy groups in the resulting spectrum.

^1H chemical shifts (ppm) and multiplicities

Coupling constants (Hz)
Ring A: $^3J_{AB} = 9.0$; $^4J_{AX} = 3.0$; Ring B: $^3J_{AX} = ^3J_{A'X} \leq 8.5$

8 Catechol (3,5,7,3',4'-pentahydroxyflavane), $C_{15}H_{14}O_6$

Sesquiterpenes and flavonoids (flavones, flavanones, flavanes) are two classes of natural substances which occur frequently in plants and which have 15 C atoms in their framework. The nine double-bond equivalents which are contained in the empirical formula, ^1H signals in the region appropriate for shielded benzene ring protons (5.9–$6.9\ ppm$) and phenolic OH protons (7.9–$8.3\ ppm$) indicate a flavonoid.

In the ^1H NMR spectrum five protons can be exchanged by deuterium. Here the molecular formula permits only OH groups. The shift values (above $7.9\ ppm$) identify four phenolic OH groups and one less acidic alcoholic OH function ($4\ ppm$, overlapping).

Between 5.8 and $6.1\ ppm$ the ^1H signals appear with typical *ortho* and *meta* couplings. The small shift values show that the benzene rings are substituted by electron donors (OH groups). In this region two subspectra can be discerned: an AB system with a *meta* coupling ($2.2\ Hz$) identifies a tetrasubstituted benzene ring A with *meta* H atoms. An

ABM system with one *ortho* and one *meta* coupling (*8.1* and *1.9 Hz*, respectively) indicates a second benzene ring **B** with a 1,2,4-arrangement of the *H* atoms. Eight of the nine double-bond equivalents are thus assigned places.

A

5.88 H^A

6.03 ^{B}H

$^4J_{AB} = 2.2 \; Hz$

B

6.76 H^B

6.79 H^A

H^M 6.89 ppm

$^3J_{AB} = 8.1 \; Hz$
$^4J_{AM} = 1.9 \; Hz$

Following the principle that the coupling partner will have the same coupling constant, one can identify in the aliphatic region a C_3 chain as a further part structure, **C**.

C

2.54 ppm d 16.0 Hz d 8.3 Hz H

2.91 ppm d 16.0 Hz d 5.0 Hz H

H 4.56 ppm d 8.3 Hz

H 4.00 ppm "t" 8.3 Hz d 5.0 Hz

Assembly of fragments **A–C**, taking into account the ninth double-bond equivalent, leads to the 3,5,7,3′,4′-pentahydroxyflavane skeleton **D** and to the following assignment of ^1H chemical shifts (*ppm*):

D

H 6.79

6.76 H

5.88 H

4.56 H

HO

H 6.89

6.03 H

H 4.00

OH H H

2.54 AB 2.91

The relative configurations of phenyl ring *B* and the O*H* groups on ring *C* follows from the *antiperiplanar* coupling (*8.3 Hz*) of the proton at *4.56 ppm*. The coupling partner *3-H* at *4.0 ppm* shows this coupling a second time (pseudotriplet 't' with *8.3 Hz* of doublets *d* with *5.0 Hz*), because one of the neighbouring methylene protons is also located in a position which is *antiperiplanar* relative to *3-H* (*8.3 Hz*) and another is located *syn* relative to *3-H* (*5.0 Hz*). Hence one concludes that this compound is catechol or its enantiomer. The stereoformula **E** shows those coupling constants (Hz) which are of significance for deriving the relative configuration.

9 Methyloxirane and monordene

The relationship $(^3J_{cis} > {}^3J_{trans}$; cf. problem *I*), which applies to cyclopropane, also holds for the *vicinal* couplings of the oxirane protons (spectrum **9a**) with the exception that here values are smaller owing to the electronegative ring oxygen atoms. As spectrum **9a** shows, the *cis* coupling has a value of *3.9 Hz* whereas the *trans* coupling has a value of *2.6 Hz*. The proton at *2.84 ppm* is thus located *cis* relative to the proton at *2.58 ppm* and *trans* relative to the proton at *2.28 ppm*. The coupling partners can be identified by their identical coupling constants where these can be read off precisely enough. Thus the methyl protons couple with the *vicinal* ring protons (*5.1 Hz*), the *cis* ring protons (*0.4 Hz*) and the *trans* ring protons (*0.5 Hz*); however, the small difference between these long-range couplings cannot be resolved in the methyl signal because of the large half-width, so that what one observes is a pseudotriplet.

Again following the principle that the same coupling constant holds for the coupling partner, the ^1H shift values (*ppm*) of the protons on the positions C-1 to C-9 of monordene can be assigned (**A**), as can the multiplicities and the coupling constants (*Hz*) (**B**).

The *relative configurations* of *vicinal* protons follow from the characteristic values of their coupling constants. Thus *16.1 Hz* confirms the *trans* relationship of the protons on C-8 and C-9, *10.8 Hz* confirms the *cis* relationship of the protons on C-6 and C-7. The *2.0 Hz* coupling is common to the oxirane protons at *3.00* and *3.27 ppm*; this value fixes the *trans* relationship of the protons at C-4 and C-5 following a comparison with the corresponding coupling in the methyloxirane (*2.6 Hz*). The *anti* relationship of the protons 4-*H* and 3-*H^A* can be recognised from their *8.7 Hz* coupling in contrast to the *syn* relationship between 3-*H^B* and 4-*H* (*3.1 Hz*). Coupling constants which are almost equal in value (*3.2–3.7 Hz*) linking 2-*H* with the protons 3-*H^A* and 3-*H^B* indicate its *syn* relationship with these protons (3-*H^A* and 3-*H^B* straddle 2-*H*).

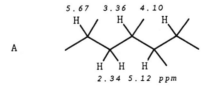

Position   3: $^3J_{AB} = 14.9$ Hz
Position 11: $^3J_{AB} = 16.3$ Hz

For larger structures the insertion of the shift values and the coupling constants in the stereo projection of the structural formula, from which one can construct a Dreiding model, proves useful in providing an overview of the stereochemical relationships.

10   2-Methyl-6-(*N,N*-dimethylamino)-*trans*-4-nitro-*trans*-5-phenylcyclohexene

An examination of the cross signals of the *HH* COSY diagram leads to the proton connectivities shown in **A** starting from the alkene proton at *5.67 ppm*.

A

Strong cross signals linking the *CH₂* group (*2.34 ppm*) with the proton at *3.36 ppm* confirm the regioselectivity of the Diels–Alder reaction and indicate the adduct **B**: the *CH₂* is bonded to the phenyl-*CH* rather than to the nitro-*CH* group; if it were bonded to the latter, then cross signals for *2.34* and *5.12 ppm* would be observed.

The stereochemistry **C** is derived from the coupling constants of the ^1H NMR spectrum: the *11.9 Hz* coupling of the phenyl-*CH* proton (*3.36 ppm*) proves its *antiperiplanar* relationship to the nitro-*CH* proton (*5.12 ppm*). In its doublet of doublets signal a second *antiperiplanar* coupling of *9.2 Hz* appears in addition to the one already mentioned, which establishes the *anti* positon of the *CH* proton at *4.10 ppm* in the positon α to the *N,N*-dimethylamino group.

Multiplicities and coupling constants (Hz)

11 (E)-3-(N,N-Dimethylamino)acrolein

First the *trans* configuration of the C-2—C-3 double bond is derived from the large coupling constant ($^3J_{HH} = 13\ Hz$) of the protons at *5.10* and *7.11 ppm*, whereby the middle *CH* proton (*5.10 ppm*) appears as a doublet of doublets on account of the additional coupling (*8.5 Hz*) to the aldehyde proton.

^1H chemical shifts (ppm)   *Multiplicities and coupling constants (Hz)*

The two methyl groups are not equivalent at 303 K (δ = *2.86* and *3.14 ppm*); rotation about the CN bond is 'frozen,' because this bond has partial π character as a result of the mesomeric effects of the dimethylamino groups (+ M) and of the aldehyde function

(− M), so that there are *cis* and *trans* methyl groups. Hence one can regard 3-(*N,N*-dimethylamino)acrolein as a vinylogue of dimethylformamide and formulate a vinylogous amide resonance.

At 318 K the methyl signals coalesce. The half-width Δv of the coalescence signal is approximately equal to the frequency separation of the methyl signals at 308 K; its value is *3.14 − 2.86 = 0.28 ppm*, which at 250 MHz corresponds to $\Delta v = 70$ Hz. The following exchange or rotation frequency of the *N,N*-dimethylamino group is calculated at the coalescence temperature:

$$k = (\pi/\sqrt{2}) \times 70 = 155.5\ \text{s}^{-1}$$

Finally from the logarithmic form of the Eyring equation, the free enthalpy of activation, ΔG, of rotation of the dimethylamino group at the coalescence temperature (318 K) can be calculated:

$$\Delta G_{318} = 19.134\ T_c \left[10.32 + \log\left(\frac{T_c\sqrt{2}}{\pi \Delta v}\right) \right] \times 10^{-3}\ \text{kJ mol}^{-1}$$

$$= 19.134 \times 3.18 \left[10.32 + \log\left(\frac{318 \times 1.414}{3.14 \times 70}\right) \right] \times 10^{-3}$$

$$= 19.134 \times 3.18\,(10.32 + 0.311) \times 10^{-3}$$

$$= 64.7\ \text{kJ mol}^{-1}\ (15.45\ \text{kcal mol}^{-1})$$

12 *cis*-1,2-Dimethylcyclohexane

The temperature dependence of the ^{13}C NMR spectrum is a result of cyclohexane ring inversion. At room temperature (298 K) four average signals are observed instead of the eight expected signals for the non-equivalent C atoms of *cis*-1,2-dimethylcyclohexane. Below −20 °C ring inversion occurs much more slowly and at −50 °C the eight expected signals of the conformers I and II appear.

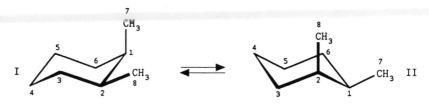

The coalescence temperatures lie between 243 and 253 K and increase as the frequency difference between the coalescing signals in the 'frozen' state increases. Thus the coalescence temperature for the pairs of signals at 35.2/33.3 ppm lies between 238 and 243 K; owing to signal overlap the coalescence point cannot be detected precisely here. The methyl signals at 20.5 and 11.5 ppm have a larger frequency difference (9 ppm or 900 Hz at 100 MHz) and so coalesce at 253 K, a fact which can be recognised from the plateau profile of the average signal (16.4 ppm). Since the frequency difference of this signal (900 Hz) in the 'frozen' state (223 K) may be measured more precisely than the width at half-height of the coalescing signal at 253 K, the exchange frequency k of the methyl groups is calculated from the equation

$$k = (\pi/\sqrt{2}) \times 900 = 1998.6 \, \text{s}^{-1}$$

The free enthalpy of activation, ΔG, of the ring inversion at 253 K is calculated from the logarithmic form of the Eyring equation:

$$\Delta G_{253} = 19.134 \, T_c \left[10.32 + \log \left(\frac{T_c \sqrt{2}}{\pi \Delta \nu} \right) \right] \times 10^{-3} \, \text{kJ mol}^{-1}$$

$$= 19.134 \times 253 \left[10.32 + \log \left(\frac{253 \times 1.414}{3.14 \times 900} \right) \right] \times 10^{-3}$$

$$= 19.134 \times 253 \, (10.32 - 1.9) \times 10^{-3}$$

$$= 40.8 \, \text{kJ mol}^{-1} \, (9.75 \, \text{kcal mol}^{-1})$$

The assignment of resonances in Table 12.2 results from summation of substituent effects as listed in Table 12.1. The data refer to conformer I; for conformer II the C atoms pairs C-1–C-2, C-3–C-6, C-4–C-5 and C-7–C-8 change places.

Table 12.1 Prediction of ^{13}C chemical shift of *cis*-1,2-dimethylcyclohexane in the 'frozen' state, using the cyclohexane shift of 27.6 ppm and substituent effects (Ref. 6, p. 316)

C-1	C-2	C-3	C-4	C-5	C-6
27.6	27.6	27.6	27.6	27.6	27.6
+ 1.4 αa	+ 6.0 αe	+ 9.0 βe	+ 0.0 γe	− 6.4 γa	+ 5.4 βa
+ 9.0 βe	+ 5.4 βa	− 6.4 γa	− 0.1 δa	− 0.2 δe	+ 0.0 γe
− 3.4 $\alpha a \beta e$	− 2.9 $\beta a \alpha e$	− 0.8 $\beta \gamma a$	+ 0.0 $\gamma e \delta e$	+ 0.0 $\gamma a \delta e$	+ 1.6 $\beta a \gamma e$
34.6	36.1	29.4	27.5	21.0	34.6

Table 12.2 Assignment of the ^{13}C resonances of *cis*-1,2-dimethylcyclohexane

Position	Predicted shift (ppm) (ppm)	Observed shift: CD$_2$Cl$_2$,223 K (ppm)	Observed shift: CD$_2$Cl$_2$298 K (ppm)
C-1	34.6	33.3	34.9
C-2	36.1	35.2	34.9
C-3	29.4	27.1	31.9
C-6	34.6	33.8	31.9
C-4	27.5	28.6	24.2
C-5	21.0	20.1	24.2
C-7 ax.	—	11.5	16.4
C-8 eq.	—	20.5	16.4

13   5-Ethynyl-2-methylpyridine

The ^{13}C NMR spectrum illustrates the connection between carbon hybridisation and ^{13}C shift on the one hand and J_{CH} coupling constants on the other.

The compound clearly contains a methyl group (24.4 ppm, quartet, $J_{CH} = 127.5$ Hz, sp³) and an ethynyl group (80.4 ppm, doublet, $J_{CH} = 252.7$ Hz, sp; 80.8 ppm, a doublet as a result of the coupling $^2J_{CH} = 47.0$ Hz). Of the five signals in the sp² shift range, three belong to CH units and two to quaternary carbon atoms on the basis of the $^1J_{CH}$ splitting (three doublets, two singlets). The coupling constant $J_{CH} = 182.5$ Hz for the doublet centred at 152.2 ppm therefore indicates a disubstituted pyridine ring **A** with a CH unit in one α-position. It follows from the shift of the quaternary C atom that the methyl group occupies the other α-position (158.4 ppm, α-increment of a methyl group, about 9 ppm, on the α-C atom of a pyridine ring, approximately 150 ppm); the shielding ethynyl group occupies a β-position, as can be seen from the small shift of the second quaternary C atom (116.4 ppm). From this, two structures **B** and **C** appear possible.

The additional doublet splitting (2.4 Hz) of the methyl quartet decides in favour of **C**; long range coupling in **B** ($^4J_{CH}$, $^5J_{CH}$) of the methyl C atom to H atoms of the ethynyl group and of the pyridine ring would not have been resolved in the spectrum. The long-range quartet splitting of the pyridine CH signal at 122.7 ppm (C-3, $^3J_{CH} = 3.7$ Hz) confirms the 2-position of the methyl group and thus locates the ethynyl group in the 5-position, as in **C**.

CH multiplicities, CH couplings (Hz), coupling *protons*:

C-2	S		d	10.5	(6-H)	d	4.3	(4-H)	d	2.4	(3-H)
C-3	D	163.6	q	3.7	(CH₃)	d	1.8	(4-H)			
C-4	D	165.0	d	5.5	(6-H)	d	1.8	(3-H)			
C-5	S		d	4.3	(3-H)	d	4.3	(6-H)	d	4.3	(β-H) ('q')
C-6	D	182.5	d	5.5	(4-H)	d	1.8	(β-H)			
2-CH₃	Q	127.5	d	2.4	(3-H)						
5-α	S		d	47.0	(β-H)						
5-β	D	252.7									

14   5-Hydroxy-3-methyl-1H-pyrazole

The compound referred to as 3-methylpyrazolone **A** ought to show a quartet and a triplet in the aliphatic region, the former for the ring CH_2 group. However, only a quartet is observed in the sp³ shift range in hexadeuteriodimethyl sulphoxide whilst at 89.2 ppm a doublet is found with $J_{CH} = 174.6$ Hz. An sp²-hybridised C atom with two cooperating $+ M$ effects fits the latter, the effect which OH and ring NH groups have in 5-hydroxy-3-methyl-1H-pyrazole **B**. The very strong shielding (89.2 ppm) could not be explained by NH tautomer C, which would otherwise be equally viable; in this case only a $+ M$ effect of the ring NH group would have any influence.

^{13}C chemical shifts (ppm)

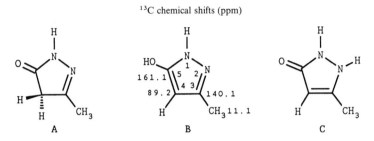

CH multiplicities, CH couplings (Hz), coupling *protons*:

C-3	S		d	6.7	(4-H)	q	6.7	(CH₃)	('qui')
C-4	D	174.6	q	3.7	(CH₃)				
C-5	S		d	3.0	(4-H)				
CH₃	Q	128.1							

15 o-Hydroxyacetophenone, C₈H₈O₂

The compound contains five double-bond equivalents. In the ^{13}C NMR spectrum all eight C atoms of the molecular formula are apparent, as a CH₃ quartet (26.6 ppm) four CH doublets (118–136 ppm) and three singlets (120.0, 162.2, 204.9 ppm) for three quaternary C atoms. The sum of these fragments (CH₃ + C₄H₄ + C₃ = C₈H₇) gives only seven H atoms which are bonded to C; since the molecular formula only contains oxygen as a heteroatom, the additional eighth H atom belongs to an OH group.

Since two quaternary atoms and four CH atoms appear in the ^{13}C NMR spectrum, the latter with a benzenoid $^3J_{CH}$ coupling constant of 7– 9 Hz, this is a disubstituted benzene ring, and the C signal with 162.2 ppm fits a phenoxy C atom. The keto carbonyl (204.9 ppm) and methyl (26.6 ppm) resonances therefore point to an acetyl group as the only meaningful second substituent. Accordingly, it must be either *o*- or *m*-hydroxyacetophenone **A** or **B**; the *para* isomer would show only four aromatic ^{13}C signals because of the molecular symmetry.

It would be possible to decide between these two by means of substituent effects, but in this case a conclusive decision is reached using the $^3J_{CH}$ coupling: the C atom marked with an asterisk in **B** would show no $^3J_{CH}$ coupling, because the *meta* positions are substituted. In the coupled ^{13}C NMR spectrum, however, all of the CH signals are split with $^3J_{CH}$ couplings of 7–9 Hz. The $^3J_{CH}$ pseudotriplet splitting of the resonance at 118.2 ppm argues in favour of **A**; the origin of the additional $^3J_{CH}$ coupling of the C atom marked with an asterisk in **A** is the intramolecular hydrogen bonding proton. This coupling also permits straightforward assignment of the closely spaced signals at 118.2 ppm (C-3) and 119.2 ppm (C-5).

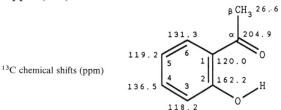

^{13}C chemical shifts (ppm)

CH multiplicities, *CH* couplings (Hz), coupling *protons*:

C-1	S		m					
C-2	S		m					
C-3	D	166.6	d	7.0	(5-H)	d	7.0	(OH)	('t')
C-4	D	161.1	d	9.1	(6-H)			
C-5	D	165.4	d	7.9	(3-H)			
C-6	D	160.8	d	8.0	(4-H)			
C-α	Q	128.1						
C-β	S		q	5.5	(CH₃)	d	5.5	(6-H)	('qui')

16   Potassium 1-acetonyl-2,4,6-trinitrophenylcyclohexadienate

The ^{13}C NMR spectrum shows from the signals at 205.6 (singlet), 47.0 (triplet) and 29.8 ppm (quartet) that the acetonyl residue with the carbonyl group intact (205.6 ppm) is bonded to the trinitrophenyl ring. Only three of the four signals which are expected for the trinitrophenyl ring from the molecular symmetry (C-1, C-2,6, C-3,5, C-4) are found here (133.4, 127.6, 121.6 ppm); however, a further doublet signal (34.5 ppm with $J_{CH} = 145.6$ Hz) appears in the aliphatic shift region. This shows that the benzene *CH* unit rehybridises from trigonal (sp²) to tetrahedral (sp³), so that a Meisenheimer salt A is produced.

Signal assignment is then no problem; the C atoms which are bonded to the nitro groups C-2,6 and C-4 are clearly distinguishable in the ^{13}C NMR spectrum by the intensities of their signals.

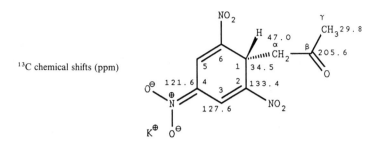

¹³C chemical shifts (ppm)

CH multiplicities, CH couplings (Hz), coupling *protons*:

C-1	D	145.6	t	4.5	*(3,5-H₂)*				
C-2,6	S		m						
C-3,5	D	166.2	d	4.4	*(1-H)*	d	4.4	*(5/3-H)*	
C-4	S		m						
C-α	T	130.1							
C-β	S		d	5.9	*(1-H)*	t	5.9	*(α-H₂)*	q 5.9 *(γ-H₃)* ('sep')
C-γ	Q	127.2							

17   *trans*-3-[4-(*N*,*N*-Dimethylamino)phenyl]-2-ethylpropenal

The relative configuration at the CC double bond can be derived from the $^3J_{CH}$ coupling of the aldehyde ^{13}C signal at 195.5 ppm in the coupled ^{13}C NMR spectrum; as a result of this coupling, a doublet (with 11.0 Hz) of triplets (with 4.9 Hz) is observed. The 11.0 Hz coupling points to a *cis* configuration of aldehyde C and alkene *H*; the corresponding *trans* coupling would have a value of *ca* 15 Hz (reference substance: methacrolein, Table 2.11). The aldehyde and *p*-dimethylaminophenyl groups therefore occupy *trans* positions.

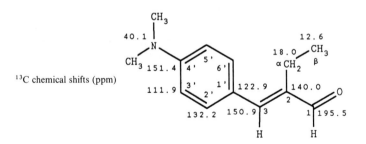

^{13}C chemical shifts (ppm)

CH multiplicites, CH couplings (Hz), coupling *protons*:

C-1	D	170.9	d	11.0	*(3-H)*	t	4.9	*(α-H₂)*	
C-2	S		d	22.0	*(1-H)*				
C-3	D	147.5	t	4.3	*(α-H₂)*	t	4.9	*(2', 6'-H₂)*	('qui')
C-1'	S		t	7.3	*(3', 5'-H₂)*				
C-2', 6'	D	158.1	d	6.7	*(3-H)*	d	6.7	*(6'/2'-H)*	('t')
C-3', 5'	D	159.3	d	5.5	*(5'/3'-H)*				
C-4'	S		m						
C-α	T	133.1	m						
C-β	Q	126.9	t	3.7	*(α-H₂)*				
N(CH₃)₂	Q	136.1	q	4.3	*(NCH₃)*				

18   *N*-Butylsalicylaldimine

In the ^{13}C NMR spectrum the signal of the *O*-trimethylsilyl group is missing near 0 ppm. Instead there is a doublet ($^1J_{CH}$ = 159.3 Hz) of quartets ($^3J_{CH}$ = 6.1 Hz) at 164.7 ppm for an imino C atom and a triplet of multiplets at 59.0 ppm. Its $^1J_{CH}$ coupling of 141.6 Hz points to an *N*-CH$_2$ unit as part of an *n*-butyl group with further signals that would fit this arrangement at 33.0, 20.3 and 13.7 ppm. The product of the reaction is therefore salicylaldehyde *N*-(*n*-butyl)imine. Assignment of the individual shifts for the hydrocarbon pair C-3–C-5, which are shielded by the hydroxy group as a +*M* substituent in the *ortho* and *para* position, respectively, is achieved by observing the possible long-range couplings: C-6 couples with 1.8 Hz to 7-*H*; C-3 is broadened as a result of coupling with the *H*-bonding OH. Both the latter and the *cis* coupling of C-α with 7-4 (7.3 Hz) point to the *E*-configuration of *N*-butyl and phenyl relative to the imino double bond.

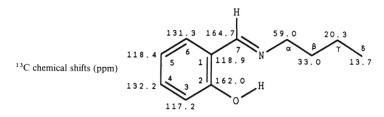

^{13}C chemical shifts (ppm)

CH multiplicities, *CH* couplings (Hz), coupling *protons*:

C-1	S		d	5.5	(*3-H*)	d	5.5	(*5-H*)	d	5.5	(*7-H*)	('q')
C-2	S		d	7.9	(*4-H*)	d	7.9	(*6-H*)	d	7.9	(*7-H*)	('q')
C-3	D	160.5	d	6.7	(*5-H*)	d	(b)	(*OH*)				
C-4	D	159.9	d	8.5	(*6-H*)	d	1.2	(*3/5-H*)				
C-5	D	163.3	d	7.6	(*3-H*)							
C-6	D	157.9	d	7.9	(*4-H*)	d	1.8	(*7-H*)				
C-7	D	159.3	d	6.1	(*4-H*)	t	6.1	(*α-H₂*)			('q')	
C-α	T	141.6	d	7.3	(*7-H*)	qui	3.4	(*β-H₂, γ-H₂*)				
C-β	T	127.6	m									
C-γ	T	127.0	m									
C-δ	Q	126.3	qui	3.1	(*β-H₂, γH₂*)							

19   Benzo[*b*]furan

All ^{13}C signals appear in the region appropriate for sp^2-hybridised C atoms; hence it could be an aromatic, a heteroaromatic or a polyene. If the matching doublet signals (·.) at the corners of the correlation square of the INADEQUATE experiment are connected (**A**) then the result is eight CC bonds, six of which relate to the benzene ring. For example, one can begin with the signal at 107.3 ppm and deduce the hydrocarbon skeleton **A**.

```
        123.6 - 122.0 - 128.4 - 107.3
A         |               |       |          ≡
        125.1 - 112.1 - 155.9   145.7
                 ↓       ↓
```

```
              122.0
                 ╱╲ 128.4   107.3
        123.6  ╱    ╲
                        145.7
        125.1  ╲    ╱
                 ╲╱  155.9
              112.1
```

The coupled ^{13}C NMR spectrum identifies the C atoms at 145.7 and 155.9 ppm as C*H* and C, respectively, whose as yet unattached bonds go to an electron-withdrawing heteroatom which causes the large shift values. The C*H* signals which are not benzenoid, at 107.3 and 145.7 ppm, show remarkably large coupling constants (177.2 and 201.7 Hz, respectively) and long-range couplings (12.7 and 11.6 Hz). These data are consistent with a 2,3-disubstituted furan ring (Tables 2.6 and 2.7); benzo[*b*]furan **B** is therefore the result.

^{13}C chemical shifts (ppm)

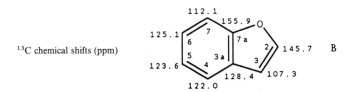

C*H* multiplicities, C*H* couplings (Hz), coupling *protons*:

C-2	D	201.7	d	11.6	(*3-H*)			
C-3	D	177.2	d	12.7	(*2-H*)	d	3.0	(*4-H*)
C-3a	S		m					
C-4	D	163.0	d	7.4	(*6-H*)	m		
C-5	D	162.3	d	8.9	(*7-H*)	m		
C-6	D	159.2	d	7.4	(*4-H*)			
C-7	D	160.0	d	6.7	(*5-H*)	d	1.5	(*6-H*)
C-7a	S		m					

Additional CC correlation signals (145.7 to 122.0; 128.4 to 125.1; 122.0 to 112.1 ppm) are the result of $^3J_{CC}$ coupling and confirm the assignments given above.

20   3-Hydroxypropyl 2-ethylcyclohexa-1,3-diene-5-carboxylate

The cross peaks in the INADEQUATE plot show the CC bonds for two part structures **A** and **B**. Taking the ^{13}C signal at 174.1 ppm as the starting point the hydrogen skeleton A and additional C$_3$ chain **B** result.

$$12.8\text{–}28.3\text{–}137.4\text{–}128.2\text{–}124.2$$
$$118.4\text{–}\ 25.4\text{–}\ 39.9\text{–}174.1 \rightarrow\ \ \leftarrow 62.1\text{–}32.0\text{–}58.6 \rightarrow$$

$$\textbf{A}\ (\text{ppm}) \qquad\qquad\qquad \textbf{B}\ (\text{ppm})$$

Part structure **A** is recognised to be a 2,5-disubstituted cyclohexa-1,3-diene on the basis of its chemical shift values. The ethyl group is one substituent, the other is a carboxy function judging by the chemical shift value of 174.1 ppm. The C*H* multiplicities which follow from the DEPT subspectra, 2C, 4C*H*, 5C*H*$_2$ and C*H*$_3$, lead to the C*H* part formula $C_2 + C_4H_4 + C_5H_{10} + CH_3 = C_{12}H_{17}$. Comparison with the given molecular formula, $C_{12}H_{18}O_3$, indicates an O*H* group. Since the C atoms at 62.1 and 58.6 ppm are linked to oxygen according to their shift values and according to the molecular formula,

A and B can be added together to form 3-hydroxypropyl 2-ethylcyclohexa-1,3-diene-5-carboxylate, **C**.

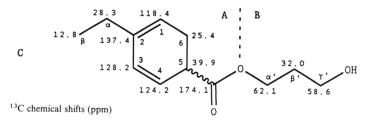

^{13}C chemical shifts (ppm)

The assignment of C-α' and C-γ' is based on the larger deshielding of C-α' by the two β-C atoms (C-γ' and C=O).

21 2-(*N,N*-Diethylamino)ethyl 4-aminobenzoate hydrochloride (procaine hydrochloride)

The discussion centres on the two structural formulae **A** and **B**.

A choice can be made between these two with the help of published ^{13}C substituent effects[5, 6] Z_i for the substituents (—NH$_2$, —NH$_3^+$, —COOR; see Section 2.5.4) on the benzene ring in **A** and **B**:

Substituent	Z_1	Z_o	Z_m	Z_p ppm
—NH$_2$	18.2	−13.4	0.8	−10.0
—NH$_3^+$	0.1	−5.8	2.2	2.2
—CO$_2$C$_2$H$_5$	2.1	−1.0	−0.5	−3.9

Adding these substituent effects gives the following calculated shift values (as compared with the observed values in parentheses) for C-1 to C-6 of the *para*-disubstituted benzene ring in **A** and **B**:

A
$$\delta_1 = 128.5 + 2.1 + 2.2 = 132.8$$
$$\delta_2 = 128.5 + 1.0 + 2.2 = 131.7$$
$$\delta_3 = 128.5 - 0.5 - 5.8 = 122.2$$
$$\delta_4 = 128.5 + 3.9 + 0.1 = 132.5 \text{ ppm}$$

B
$$\delta_1 = 128.5 + 2.1 - 10.0 = 120.6 \quad (115.5)$$
$$\delta_2 = 128.5 + 1.0 + 0.8 = 130.3 \quad (131.5)$$
$$\delta_3 = 128.5 - 0.5 - 13.4 = 114.6 \quad (113.1)$$
$$\delta_4 = 128.5 + 3.9 + 18.2 = 150.6 \text{ ppm} \quad (153.7 \text{ ppm})$$

Substituent effects calculated for structure **B** lead to values which are not perfect but which agree more closely than for **A** with the measured ^{13}C shifts of the benzene ring carbon atoms. The diastereotopism of the NCH$_2$ protons in the ^1H NMR spectrum also points to **B** as the Newman projection along the CH$_2$— ammonium-N bond shows:

Hence one finds two overlapping pseudotriplets (*3.41* and *3.44 ppm*) for the NCH$_2$ group which appears only once and two overlapping quartets (*3.22* and *3.25 ppm*) for the NCH$_2$ groups which appear twice. Since the shift differences of the CH$_2$ protons are so small, the expected *AB* system of the coupling partner approximates to an A_2 system; thus one observes only the central multiplet signals of this *AB* system.

The assignment of the ^{13}C NMR spectrum is based on the different $^1J_{CH}$ coupling constants of OCH$_2$ (149.4 Hz) and NCH$_2$ groups (140–142 Hz). With benzenoid $^3J_{CH}$ couplings the influence of the different electronegativities of the substituents on the coupling path (4.5 Hz for NH$_2$ and 6.6 Hz for COOR) and on the coupling C atom is very obvious (8.8 Hz for NH$_2$ at C-4 and 7.7 Hz for COOR at C-1).

Chemical shifts (ppm, ^{13}C: upright; ^1H: *italics*)

CH multiplicities, CH couplings (Hz), coupling *protons*:

C-1	S		t	7.7	(3, 5-H$_2$)
C-2, 6	D	158.9	d	6.6	(6/2-H)
C-3, 5	D	158.9	d	4.5	(5/3-H)
C-4	S		t	8.8	(2, 6-H$_2$)
COO	S		m		
C-α	T	149.4			
C-β	T	142.0	m		
C-α'	T	140.0	m		
C-β	Q	128.5			

HH coupling constants (Hz):
$^3J_{AX} \geq 8.6$; $^3J_{\alpha\beta} = 5.0$; $^3J_{\alpha'\beta'} = 5.0$

22   2-Ethoxycarbonyl-4-(3-hydroxypropyl)-1-methylpyrrole

Here it is possible to consider how the starting materials may react and to check the result with the help of the spectra. Another approach would start by tabulating the ^{13}C shifts, CH multiplicities and CH coupling constants and where possible the ^1H shifts and

Table 22.1 Interpretation of the NMR spectra in **22**

No.	δ_c (ppm)		J_{CH} (Hz)	δ_H (ppm)	J_{HH} (Hz)	
1	161.2	S	CO(O)			
2	127.5	D	CH	181.6	6.73	d	1.9
3	122.8	S	C			
4	121.9	S	C			
5	117.0	D	CH	172.2	6.52	d	1.9
6	61.7	T	OCH_2	140.7	3.57	t	7.0
7	59.5	T	OCH_2	147.1	4.18	q	7.2
8	36.3	Q	NCH_3	140.2	3.78	s	
9	33.5	T	CH_2	126.4	1.74	qui	6.9
10	22.5	T	CH_2	126.3	2.43	t	7.0
11	14.2	Q	CH_3	126.7	1.26	t	7.2
CHpartial formula			$C_{11}H_{16}(NO_3)$			

the HH coupling constants (Table 22.1). From this it is possible to identify those parts of the starting materials that have remained intact and those which have been lost and also those H atoms which are linked to carbon and to heteroatoms.

   This evaluation reveals that the three substructures of the reagents that are also present in the product include the N-methyl group (signals 8), the ethoxycarbonyl group (signals 1, 7, 11) and the n-propyloxy group of the dihydro-2H-pyran ring (signals 6, 9, 10). The ethyl ester OCH_2 group can also be identified in the ^{13}C NMR spectrum because of its long-range quartet splitting (4.5 Hz). The H atom missing in the CH balance but present in the molecular formula appears in the ^1H NMR spectrum as a broad D_2O-exchangeable signal (*3.03 ppm*); since the compound only contains one N atom in the form of an NCH_3 group, the signal at *3.03 ppm* must belong to an OH group. Hence the dihydro-2H-pyran ring has opened.

   By contrast, the aldehyde signals of the reagent *1* are missing from the NMR spectra. Instead an AB system appears in the ^1H NMR spectrum (*6.52* and *6.73 ppm* with $J_{AB} = 1.9$ Hz) whilst in the ^{13}C NMR spectrum two doublets appear (117.0 and 127.5 ppm) as well as two singlets (121.9 and 122.8 ppm), of which one doublet (127.5 ppm) is notable for the fact that it has a large CH coupling constant (181.6 Hz). This value fits the α-C atom of an enamine fragment (for the α-C of an enol ether fragment $J_{CH} \gtrsim 190$ Hz would be expected). This leads to a 1,2,4-trisubstituted pyrrole ring **5**, given the three double-bond equivalents (the fourth has already been assigned to the carboxy group), the AB system in the ^1H NMR spectrum (*6.52, 6.73 ppm*), the N-methyl group (signal 8) and the four ^{13}C signals in the sp^3 shift range (117–127.5 ppm). The formation of the ring from reagents *1* and *2* via intermediates *3* and *4* can be inferred with no difficulty. All ^1H and ^{13}C signals can be identified without further experiment by using their shift values, multiplicities and coupling constants.

Chemical shifts (ppm, ^{13}C: upright; ^1H: *italics*)

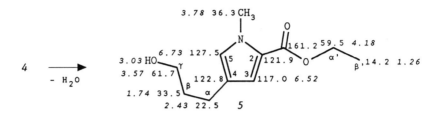

$4 \xrightarrow{-H_2O}$

CH multiplicities, *CH* couplings (Hz), coupling *protons*:

C-2	S		d	6.5	(*3-H*)	d	6.5	(*5-H*)	('t')	
C-3	D	172.2	d	5.0	(*5-H*)	t	5.0	(*6-H₂*)	('q')	
C-4			m							
C-5	D	181.6	d	7.0	(*3-H*)	t	3.9	(*6-H₂*)		
NCH₃	Q	140.2	s							
COO	S		b							
C-α	T	126.3	m							
C-β	T	126.4	m							
C-γ	T	140.7	t	4.4	(*6-H₂*)	t	4.4	*7-H₂*)	('qui')	
C-α'	T	147.1	q	4.5	(*β'-H₃*)					
C-β'	Q	126.7	b							

HH coupling constants (Hz):
$^3J_{3,5} = 1.9$; $^3J_{\alpha,\beta} = 7.0$; $^3J_{\beta,\gamma} = 7.0$; $^3J_{\alpha',\beta'} = 7.2$

23   2-*p*-Tolylsulphonyl-5-propylpyridine

The NMR spectra show that the product of the reaction contains:

—the propyl group **A** of 1-ethoxy-2-propylbuta-1,3-diene,

A

$$
\begin{array}{cc}
 & 1.53\ sxt \\
 & 23.8\ T \\
13.4\ Q & 34.6\ T \\
0.82\ t & 2.55\ t
\end{array}
$$

—the *p*-tolyl residue **B** from *p*-toluenesulphonyl cyanide.

B

$$
\begin{array}{cc}
7.75\ XX' & 7.20\ AA' \\
128.6\ Dd & 129.6\ Dqui \\
136.2\ t & 141.9\ b \\
-O_2S- & -CH_3\ \ 2.28\ s \\
 & 21.5\ Q
\end{array}
$$

—and (on the basis of their typical shift values and coupling constants, e.g. $J_{CH} = 180.2$ Hz at 150.5 ppm), a disubstituted pyridine ring **C** (three ^1H signals in the ^1H NMR, three *CH* doublets in the ^{13}C NMR spectrum) with substituents in the 2- and 5-positions, because in the ^1H NMR spectrum the $8.2\,Hz$ coupling appears instead of the $5\,Hz$ coupling ($^3J_{AB} = {}^3J_{3\text{-}H,\,4\text{-}H}$),

However, the ethoxy group of 1-ethoxy-2-propylbuta-1,3-diene is no longer present.

Evidently the *p*-toluensulphonyl cyanide (2) undergoes cycloaddition to 1-ethoxy-2-propylbuta-1,3-diene (1). The resulting dihydropyridine 3 aromatises with 1,4-elimination of ethanol to form 2-*p*-tolylsulphonyl-5-propylpyridine (4). Complete assignment is possible without further experiments using the characteristic shifts, multiplicities and coupling constants.

Chemical shifts (ppm, ^{13}C: upright; ^1H: *italics*)

CH multiplicities, *CH* couplings (Hz), coupling *protons*:

C-2	S		d	11.8	(6-*H*)	d	8.9	(4-*H*)	
C-3	D	170.3							
C-4	D	163.4	d	4.9	(6-*H*)	t	4.9	(α-*H₂*)	('q')
C-5	S		d	10.0	(6-*H*)	d	5.0*	(3-*H*)	t*5.0 (β-*H₂*) *('q')
C-6	D	180.2	d	5.9	(4-*H*)	t	5.9	(α-*H₂*)	('q')
C-α	T	127.0	b						
C-β	T	127.0	t	4.9	(α-*H₂*)	q	4.9	(γ-*H₃*)	('sxt')
C-γ	Q	126.2	t	3.9	(β-*H₂*)				
C-1'	S		t	8.9	(3',5'-*H₂*)				
C-2',6'	D	166.4	d	6.0	(6'/2'-*H*)				
C-3',5'	D	161.5	d	6.0	(5'/3'-*H*)	q	6.0	(CH₃)	('qui')
C-4	S		m						
4'-CH₃	Q	127.0	t	4.4	(3',5'-*H₂*)				

HH coupling constants (Hz):
$^3J_{AM} = 8.2;\ {}^4J_{MX} = 1.8;\ {}^3J_{AX(A'X')} \leqslant 8.4;\ {}^3J_{\alpha\beta} = 7.3;\ {}^3J_{\beta\gamma} = 7.3$

24   Triazolo[1,5-*a*]pyrimidine

Without comparative data on authentic samples, ^{13}C NMR allows no differentiation between isomers *3* and *4*; ^{13}C chemical shifts and C*H* coupling constants are consistent with either isomer.

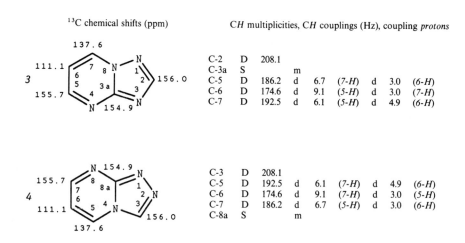

^{13}C chemical shifts (ppm)

C*H* multiplicities, C*H* couplings (Hz), coupling *protons*

C-2	D	208.1						
C-3a	S		m					
C-5	D	186.2	d	6.7	(*7-H*)	d	3.0	(*6-H*)
C-6	D	174.6	d	9.1	(*5-H*)	d	3.0	(*7-H*)
C-7	D	192.5	d	6.1	(*5-H*)	d	4.9	(*6-H*)

C-3	D	208.1						
C-5	D	192.5	d	6.1	(*7-H*)	d	4.9	(*6-H*)
C-6	D	174.6	d	9.1	(*7-H*)	d	3.0	(*5-H*)
C-7	D	186.2	d	6.7	(*5-H*)	d	3.0	(*6-H*)
C-8a	S		m					

However, in ^{15}N NMR spectra, the $^{2}J_{NH}$ coupling constants ($\geqslant 10$ Hz) are valuable criteria for structure determination. The ^{15}N NMR spectrum shows $^{2}J_{NH}$ doublets with 11.8, 12.8 and 15.7 Hz for all of the imino N atoms. Therefore, triazolo[1,5-*a*]pyrimidine (*3*) is present; for the [4,3-*a*] isomer *4*, nitrogen atom N-1 would appear as a singlet signal because it has no *H* atoms at a distance of two bonds. This assignment of the ^{15}N shifts is supported by a comparison with the spectra of derivatives which are substituted in positions 2 and 6.[8] If a substituent is in position 6 then the 1.5 Hz coupling is lost for N-4; for substitution in position 2 or 6 a doublet instead of a triplet is observed for N-8. The ^{15}N shift and the $^{2}J_{NH}$ coupling constants of N-1 are considerably larger than for N-3 as a result of the electronegativity of the neighbouring N-8.

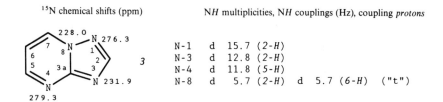

^{15}N chemical shifts (ppm)

N*H* multiplicities, N*H* couplings (Hz), coupling *protons*

N-1	d	15.7	(*2-H*)			
N-3	d	12.8	(*2-H*)			
N-4	d	11.8	(*5-H*)			
N-8	d	5.7	(*2-H*)	d	5.7	(*6-H*)	("t")

25   6-*n*-Butyltetrazolo[1,5-*a*]pyrimidine and 2-azido-5-*n*-butylpyrimidine

Tetrazolo[1,5-*a*]pyrimidine (*1*) exists in equilibrium with its valence isomer 2-azidopyrimidine (*2*).

In all types of NMR spectra (^1H, ^{13}C, ^{15}N), 2-azidopyrimidine (*2*) can be distinguished by the symmetry of its pyridine ring (chemical equivalence of *4-H* and *6-H*, C-4 and C-6, N-1 and N-3) from tetrazolo[1,5-*a*]pyrimidine (*1*) because the number of signals is reduced by one. Hence the prediction in Table 25.1 can be made about the number of resonances for the *n*-butyl derivative.

All of the NMR spectra indicate the predominance of the tetrazolo[1,5-*a*]pyrimidine *1* in the equilibrium by the larger intensity (larger integral) of almost all signals, although the non-equivalence of the outer *n*-butyl C atoms in both isomers (at 22.5 and 13.9 ppm) cannot be resolved in the ^{13}C NMR spectrum. By measuring the integrated intensities, for example, one obtains for the signals showing $^2J_{NH}$ splitting (of 12.0 and 11.5 Hz, respectively) recognisable signal pairs of the pyrimidine N atoms (*1*, at 275.6; *2*, at 267.9 ppm) of integrated intensities 30.5 and 11.0 mm. Since two N nuclei generate the signal at 267.9 ppm because of the chemical equivalence of the ring N atoms in *2* its integral must be halved (5.5 mm). Thus we obtain

$$\%2 = 100 \times 5.5/(30.5 + 5.5) = 15.3\%$$

The evaluation of other pairs of signals in the ^1H and ^{15}N NMR spectra leads to a mean value of 15.7 ± 0.5% for *2*. Therefore, 6-*n*-butyltetrazolo[1,5-*a*]pyrimidine (*1*) predominates in the equilibrium with 84.3 ± 0.5%.

Assignment of the signals is completed in Table 25.2. The criteria for assignment are the shift values (resonance effects on the electron density on C and N), multiplicities and coupling constants. Because the difference between them is so small, the assignment of N-β and N-γ is interchangeable.

Table 25.1   The number of signals from *1* and *2* in the NMR spectra

Compound	^1H signals	^{13}C signals	^{15}N signals
1	6	8	5
2	5	7	4

Table 25.2 Assignment of the signals from 6-*n*-butyltetrazolo[1, 5-*a*]pyrimidine (*1*) and 2-azido-5-*n*-butylpyrimidine (*2*)

Multiplicities, coupling constants (Hz), coupling protons:

Compound	J_{CH}	$^3J_{CH}$	$^3J_{HH}$	$^2J_{NH}R$
1 3a	S	d14.7(*5-H*)		
4				d12.0(*5-H*)
5	D 185.9	d 5.0(*7-H*) t 5.0(*α-H₂*)	*d 1.8(7-H)*	
7	D 193.1	d 5.0(*5-H*) t 5.0(*α-H₂*)	*d 1.8(5-H)*	
α	T 128.5	b	*t 7.4(β-H₂)*	
β	T 128.7	b	*qui7.4(α, γ-H₄)*	
γ	T 126.6	b	*sxt7.4(β, δ-H₅)*	
δ	Q 125.5	b	*t 7.4(γ-H₂)*	
2 1, 3				d11.5(*4/6-H*)
2	S	t12.5(*4, 6-H₂*)		
4, 6	D 180.2	d 5.0(*6/4-H*) t 5.0(*α-H₂*)		
α	T 127.7	b	*t 7.4(β-H₂)*	
β	T o		*qui7.4(α, γ-H₄)*	
δ	T o		*sxt7.4(β, δ-H₅)*	
γ	Q o		*t 7.4(γ-H₂)*	

Chemical shifts (ppm, ^{13}C: upright; ^{15}N: **bold**; ^{1}H: *italics*)

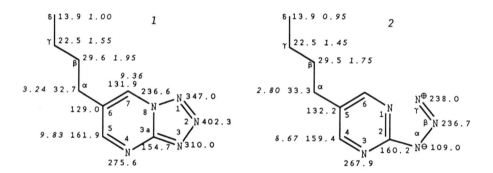

26 Hex-3-yn-1-ol, $C_6H_{10}O$

All six of the C atoms found in the molecular formula appear in the ^{13}C NMR spectrum. Interpretation of the $^1J_{CH}$ multiplets gives one CH_3 group (14.4 ppm), three CH_2 groups (12.6, 23.2 and 61.6 ppm) and two quaternary C atoms (76.6 and 83.0 ppm). The addition of these CH fragments ($CH_3 + C_3H_6 + C_2$) produces C_6H_9; the additional H atom in the molecular formula therefore belongs to an OH group. This is a part of a primary alcohol function CH_2OH, because a ^{13}C shift of 61.6 ppm and the corresponding splitting (triplet, $^1J_{CH} = 144.0$ Hz) reflect the $-I$ effect of a neighbouring O atom. The long-range triplet splitting of the CH_2O signal (6.3 Hz) indicates a neighbouring CH_2

group. This hydroxyethyl partial structure **A** is evident in the ^1H NMR spectrum also, in which the coupling proton may be identified by the uniformity of its coupling constants.

```
                    T130.9    T144.0                    Hz
                     23.2      61.6                     ppm
       A        –   CH₂   –   CH₂   –   OH
                     2.32      3.58      4.72 ppm
                     t7.1      t7.1                     Hz
                     t2.2      d4.9      t4.9           Hz
```

Obviously the exchange frequency of the OH protons is small in comparison with the coupling constant (*4.9 Hz*), so coupling between the OH and CH_2 protons also causes additional splitting of the ^1H signals (*3.58* and *4.72 ppm*).

The additional triplet splitting (*2.2 Hz*) of the CH_2 protons (at *2.32 ppm*) is the result of long-range coupling to the third CH_2 group of the molecule, which can be recognised at *2.13 ppm* by the same fine structure. The larger coupling constant (*7.6 Hz*) is repeated in the triplet at *1.07 ppm*, so that an ethyl group is seen as a second structural fragment **B** in accordance with the further signals in the ^{13}C NMR spectrum (12.6 ppm, T 130.4 Hz), and 14.4 ppm, Q 127.9 Hz, t 5.4 Hz).

```
    t   5.4   q   4.4              b        t   6.3              Hz
   Q127.9    T130.4          T130.9  T144.0                     Hz
    14.4      12.6            23.2     61.6                      ppm
    CH₃   –   CH₂ –       –  CH₂  –  CH₂  –  OH
    1.07      2.13               2.32     3.58     4.72  ppm
    t7.6      q7.6               t7.1     t7.1              Hz
              t2.2 .............. t2.2     d4.9     t4.9   Hz
         B                              A
```

The long-range coupling of 2.2 Hz which appears in **A** and **B**, two quaternary C atoms in the ^{13}C NMR spectrum with appropriate shifts (76.6 and 83.0 ppm) and the two double-bond equivalents (molecular formula) suggest that a CC triple bond links the two structural fragments. This is confirmed by the CC correlation experiment (INADEQUATE), which, apart from the triple bond itself (because the relaxation time of the C atoms in question is too long, the CC triple bond is not visible in the INADEQUATE spectrum), establishes the molecular skeleton (formula **C**). The *AB* system of the C atoms at 12.6 and 14.4 ppm, because of the small shift difference, approaches an A_2 situation, so that only the inner *AB* signals appear with sufficient intensity. Hence the compound is identified as hex-3-yn-1-ol (**C**) in accordance with the coupling patterns.

Chemical shifts (ppm, ^{13}C: upright; ^1H: *italics*)

```
        1.07  14.4  6
                          4   76.6
     C   2.13  12.6   5                      23.2  2.32
                          83.0   3       2
                              3.58  61.6       1
                                          OH 4.72
```

CH multiplicities, *CH* couplings (Hz), coupling *protons*:

C-1	T	144.0	t	6.3	(2-H_2)					
C-2	T	130.9	b							
C-3	S		t	9.0	(2-H_2)	t*4.5	(1-H_2)	t*4.5	(5-H_2)	*(qui)
C-4	S		t	10.5	(5-H_2)	q*5.0	(6-H_2)	t*5.0	(2-H_2)	*(sep)
C-5	T	130.4	q	4.4	(6-H_3)					
C-6	Q	127.9	t	5.4	(5-H_2)					

HH multiplicities, *HH* couplings (Hz), coupling *protons*:

1-H_2	t	7.1	(2-H_2)	d	4.9	(OH)	
2-H_2	t	7.1	(1-H_2)	t	2.2	(5-H_2)	
5-H_2	q	7.6	(6-H_3)	t	2.2	(2-H_2)	
6-H_3	t	7.6	(5-H_2)				
OH	t	4.9	(1-H_2)				

27 6-Methoxytetralin-1-one, $C_{11}H_{12}O_2$

Almost all parts of the structure of this compound are already apparent in the ^1H NMR spectrum. It is possible to recognise:

—three methylene groups linked to one another, **A**,

```
        t6.2   qui6.2    t6.2        Hz
        2.87    2.09     2.56       ppm
A       — CH2 —  CH2  —  CH2 —
```

—a methoxy group **B**,

```
B              — OCH3   3.81 ppm
```

—and a 1,2,4-trisubstituted benzene ring **C**, in the following way:

The signal at *6.79 ppm* splits into a doublet of doublets. The larger coupling (*8.7 Hz*) indicates a proton in the *ortho* position, the smaller (*2.5 Hz*) a further proton in a *meta* position, and in such a way that the *ortho* proton (*7.97 ppm*) does not show any additional *ortho* coupling.

The ^{13}C NMR spectrum confirms

—three methylene groups **A** (23.6, 30.3, 39.1 ppm, triplets),
—the methoxy group **B** (55.7 ppm, quartet),
—the trisubstituted benzene ring **C** (three *CH* doublets and three quaternary C atoms between 113.3 and 164.6 ppm)
—and identifies additionally a keto-carbonyl group **D** at 197.8 ppm.

Five double-bond equivalents can be recognised from the shift values (four for the benzene ring and one for the carbonyl group). The sixth double-bond equivalent implied by the molecular formula belongs to another ring, so that the following pieces can be drawn for the molecular jigsaw puzzle:

B   C   D   A

The methoxy group is a $+M$ substituent, and so shields *ortho* protons and C atoms in *ortho* positions; the protons at *6.67* and *6.79 ppm* reflect this shielding. The carbonyl group as a $-M$ substituent deshields *ortho* protons, and is *ortho* to the proton at *7.97 ppm*. With the additional double-bond equivalent for a ring, 6-methoxytetralin-1-one (**E**) results.

The difference between 2 CH_2 and 4 CH_2 is shown by the nuclear Overhauser enhancement (NOE) on the proton at *6.67 ppm*, if the methylene protons are irradiated at *2.87 ppm*. The arrangement of the methylene C atoms can be read from the *CH* COSY segment. The C atoms which are in close proximity to one another at 113.3 and 113.8 ppm belong to C-5 and C-7. Carbon atom C-5 is distinguished from C-7 by the pseudo-quartet splitting ($^3J_{CH} = 3.4$ Hz to *7-H* and *4-H$_2$*) that involves the methylene group in the *ortho* position.

Chemical shifts (ppm, ^{13}C: upright; ^1H: italics)

CH multiplicities, *CH* couplings (Hz), coupling *protons*:

C-1	S		d	8.0	*(8-H)*	t	4.0	*(2-H$_2$)*	t	4.0	*(3-H$_2$)*	('qui')
C-2	T	127.3	b									
C-3	T	129.0	t	3.4	*(2-H$_2$)*	t	3.4	*(4-H$_2$)*				('qui')
C-4	T	130.0	b									
C-4a	S		d	4.0	*(8-H)*	t	4.0	*(4-H$_2$)*				('q')
C-5	D	158.3	d	3.4	*(7-H)*	t	3.4	*(4-H$_2$)*				('q')
C-6	3		ili									
C-7	D	156.6	d	5.2	*(6-H)*							
C-8	D	161.2	s									
C-8a	S		m									
OCH$_3$	Q	144.5	s									

HH multiplicities, *HH* coupling constants *(Hz)*, coupling *protons*:

2-H$_2$	t	6.2	*(3-H$_2$)*				
3-H$_2$	t	6.2	*(2-H$_2$)*	t	6.2	*(4-H$_2$)*	('qui')
4-H$_2$	t	6.2	*(3-H$_2$)*				
5-H	d	2.5	*(7-H)*				
7-H	d	8.7	*(8-H)*	d	2.5	*(5-H)*	
8-H	d	8.7	*(7-H)*				
OCH$_3$	s						

28 Hydroxyphthalide

The ^1H NMR spectrum does not show a signal for either a carboxylic acid or an aldehyde function. Instead, a D_2O-exchangeable signal appears in the range of less acidic

OH protons (*4.8 ppm*) and a non-exchangeable signal appears at *6.65 ppm*. The latter fits a *CH* fragment of an acetal or hemiacetal function which is strongly deshielded by two O atoms, also confirmed by a doublet at *98.4 ppm* with J_{CH} = 174.6 Hz in the ^{13}C NMR spectrum. According to this it is not phthalaldehydic acid (*1*) but its acylal, hydroxyphthalide (*2*).

The conclusive assignment of the ^{1}H and ^{13}C signals of the *ortho*-disubstituted benzene ring at 80 and 20 MHz, respectively, encounters difficulties. However, the frequency dispersion is so good at 400 and 100 MHz, respectively, that the *HH* COSY in combination with the *CH* COSY technique allows a conclusive assignment to be made. Proton connectivities are derived from the *HH* COSY; the *CH* correlations assign each of the four *CH* units. Both techniques converge to establish the *CH* skeleton of the *ortho*-disubstituted benzene ring.

```
                 7.84      7.63      7.76      7.67    ppm
         A        |         |         |         |
                125.1  -  130.6  -  134.6  -  123.6    ppm
                 C-6       C-5       C-4       C-3
```

Reference to the deshielding of a ring proton by an *ortho* carboxy group clarifies the assignment.

Chemical shifts (ppm, ^{13}C: upright; ^{1}H: *italics*)

CH multiplicities, *CH* couplings (Hz), coupling *protons*:

```
C-1   S          m
C-2   S          d   7.5   (4-H)   d   7.5   (6-H)   ('t')
C-3   D   165.4  d   6.7   (5-H)
C-4   D   166.0  d   7.0   (6-H)
C-5   D   162.4  d   7.3   (3-H)
C-6   D   162.4  d   5.5   (4-H)
C-7   S          b
C-8   D   174.6  b
```

HH multiplicities, *HH* couplings (Hz), coupling protons:

```
3-H   d   7.5   (4-H)
4-H   d   7.5   (3-H)   d   7.5   (5-H)   ('t')
5-H   d   7.5   (4-H)   d   7.5   (6-H)   ('t')
6-H   d   7.5   (5-H)
```

29   Nona-2-*trans*-6-*cis*-dienal

From the *HH* COSY plot the following *HH* connectivities **A** are derived:

A   *0.92→ 1.99→ 5.39→ 5.26→ 2.22→ 2.36→ 6.80→ 6.08→ 9.45*   ppm

In the *CH* COSY plot it can be established which C atoms are linked with these protons; thus the *CH* skeleton **B** can readily be derived from **A**:

0.92→ 1.99→ 5.39→ 5.26→ 2.22→ 2.36→ 6.80→ 6.08→ 9.45   ppm

B | | | | | | | | |

14.2- 20.5-133.3-126.7- 25.4- 32.7-158.1-133.2-194.0   ppm

Structural elucidation can be completed to give **C** if the *CH* multiplicities from the ^{13}C NMR spectrum and characteristic chemical shift values from the ^{1}H and ^{13}C NMR spectra are also taken into account. The shift pair 194.0/*9.45* ppm, for example, clearly identifies an aldehyde group; the shift pairs 133.2/*6.08*, 158.1/*6.80*, 126.7/*5.26* and 133.5/*5.39* ppm identify two CC double bonds of which one (133.1/*6.08* and 158.1/*6.80* ppm) is polarised by the $-M$ effect of the aldehyde group.

```
        9       8       7   6       5       4       3   2       1
C     CH3  -  CH2  -  CH  = CH  -  CH2  -  CH2  -  CH  = CH  -  CH  = O
```

Hence the compound is nona-2,6-dienal. The relative configuration of both CC double bonds follows from the *HH* coupling constants of the alkene protons in the ^{1}H NMR spectrum. The protons of the polarised 2,3-double bond are in *trans* positions ($^{3}J_{HH} = 15.5\ Hz$) and those on the 6,7-double bond are in *cis* positions ($^{3}J_{HH} = 10.5\ Hz$). The structure is therefore nona-2-*trans*-6-*cis*-dienal, **D**.

In assigning all shift values, *CH* coupling constants and *HH* coupling constants, differentiation between C-2 and C-7 is at first difficult because the signals are too crowded in the ^{13}C NMR spectrum. Differentiation is possible, however, on closer examination of the *CH* COSY plot and the coupled ^{13}C NMR spectrum: the signal at 133.2 ppm splits as a result of *CH* long-range couplings into a doublet (25.0 Hz) of triplets (5.7 Hz), whose 'left' halves overlap in each case with the less clearly resolved long-range multiplets of the neighbouring signal, as the signal intensities show. Thereby, the coupling constant of 25.0 Hz locates the aldehyde proton which is two bonds apart from the C atom at 133.2 ppm.

Chemical shifts (ppm, ^{13}C: upright; ^{1}H: italics)

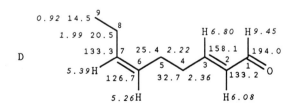

CH multiplicities, CH couplings (Hz), coupling *protons*

C-1	D	171.0	d	9.5	(2-H)							
C-2	D	160.2	o	d	25.0	(1-H)	t	5.7	(4-H₂)			
C-3	D	151.3	t	5.5	(4-H₂)	t	5.5	(5-H₂)			('qui')	
C-4	T	127.2	m									
C-5	T	126.5	m									
C-6	D	155.1	t	4.5	(4-H₂)	t	4.5	(5-H₂)	t	4.5	(8-H₂)	('sep')
C-7	D	158.3	o									
C-8	T	127.8	m									
C-9	Q	126.5	m									

HH multiplicities, HH couplings (Hz), coupling protons:

1-H	d	7.9	(2-H)								
2-H	d	15.5	(3-H, trans)	d	7.9	(1-H)			t	1.4	(4-H₂)
3-H	d	15.5	(2-H, trans)	t	6.9	(4-H₂)					
4-H₂	d	6.9	(3-H)	t	6.9	(5-H₂)	('q')				
5-H₂	d	7.0	(6-H)	t	7.0	(4-H₂)	('q')				
6-H	d	10.5	(7-H, cis)	t	7.0	(5-H₂)			t	1.2	(8-H₂)
7-H	d	10.5	(6-H, cis)	t	7.0	(8-H₂)			t	1.4	(5-H₂)
8-H₂	d	7.0	(7-H)	q	7.0	(9-H₃)	('qui')	d	1.2	(6-H)	
9-H₃	t	7.0	(8-H₂)								

30 *trans*-1-Cyclopropyl-2-methylbuta-1,3-diene (*trans*-isopren-1-yl-cyclopropane)

In the ^{13}C NMR spectrum two signals with unusually small shift values $[(CH_2)_2$: 7.5 ppm: CH: 10.6 ppm] and remarkably large CH coupling constants (161.9 and 160.1 Hz) indicate a monosubstituted cyclopropane ring **A**. The protons which belong to this structural unit at *0.41 (AA')*, *0.82 (BB')* and *1.60 ppm (M)* with typical values for *cis* couplings *(8.1 Hz)* and *trans* couplings *(4.9 Hz)* of the cyclopropane protons can be identified from the CH COSY plot.

Chemical shifts (ppm)   HH coupling constants (Hz)

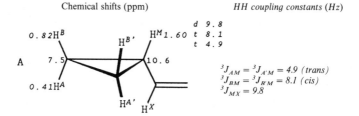

$$^3J_{AM} = {}^3J_{A'M} = 4.9 \ (trans)$$
$$^3J_{BM} = {}^3J_{B'M} = 8.1 \ (cis)$$
$$^3J_{MX} = 9.8$$

The additional coupling *(9.8 Hz)* of the cyclopropane proton *M* at *1.60 ppm* is the result of a *vicinal H* atom in the side-chain. This contains a methyl group **B**, a vinyl group **C** and an additional substituted ethenyl group **D**, as may be seen from the one-dimensional ^1H and ^{13}C NMR spectra and from the CH COSY diagram.

Since the vinyl-C*H* proton at *6.33 ppm* shows no additional $^3J_{HH}$ couplings apart from the doublet of doublets splitting (*cis* and *trans* coupling), the side-chain is a 1-isoprenyl chain **E** and not a 1-methylbuta-1,3-dienyl residue **F**.

Hence it must be either *trans*- or *cis*-1-cyclopropyl-2-methylbuta-1,3-diene (1-isoprenylcyclopropane), **G** or **H**.

In decoupling the methyl protons, the NOE difference spectrum shows a nuclear Overhauser enhancement on the cyclopropane proton at *1.60 ppm* and on the terminal vinyl proton with *trans* coupling at *5.05 ppm* and, because of the *geminal* coupling, a negative NOE on the other terminal proton at *4.87 ppm*. This confirms the *trans* configuration **G**. In the *cis* isomer **H** no NOE would be expected for the cyclopropane proton, but one would be expected for the alkenyl-*H* in the α-position indicated by arrows in **H**.

Chemical shifts (ppm, ^{13}C: upright; ^1H: italics)

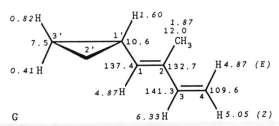

G

C*H* multiplicities, C*H* couplings (Hz), coupling *protons*:

C-1′	D	160.1				m				
C-2′3′	T	161.9				m				
C-1	D	150.4				m				
C-2	S					m				
C-3	D	151.3				d	8.0	(*1-H*)	q	4.0	(*CH₃*)
C-4	D	158.7	D	153.9						
2-C*H₃*	Q	125.6				d	8.0	(*1-H*)	d	4.4	(*3-H*)

HH multiplicities, *HH* couplings *(Hz)*, coupling *protons*

1′-H^{M*}	d	9.8	(*1-H*)	t	8.1	(2′,3′-H$^{BB'}$)	t	4.9	(2′,3′-H$^{AA'}$)
1-H	d	9.8	(*1′/H*)						
3-H	d	17.0	(*4-H E*)	d	11.0	(*4-H Z*)			
4-H (E)	d	17.0	(*3-H*)						
4-H (Z)	d	11.0	(*3-H*)						
5-C*H₃*	d	1.5	(*1′-H*)						

* The cyclopropane protons form an AA′BB′M system.

31 Dicyclopentadiene

The ^{13}C NMR spectrum does not show the three resonances expected for monomeric cyclopentadiene. Instead, ten distinct signals appear, of which the DEPT spectrum identifies four CH carbon atoms in each of the shift ranges appropriate for alkanes and alkenes and in the alkane range an additional two CH_2 carbon atoms. This fits the [4 + 2]-adduct *2* of cyclopentadiene *1*.

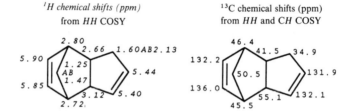

The structure of the dimer can be derived simply by evaluation of the cross signals in the *HH* COSY plot. The cycloalkene protons form two *AB* systems with such small shift differences that the cross signals lie within the contours of the diagonal signals.

^{1}H chemical shifts (ppm)
from *HH* COSY

```
      2.80
         2.66  1.60AB2.13
5.90
         1.25
         AB          5.44
5.85     1.47
            3.12  5.40
         2.72
```

^{13}C chemical shifts (ppm)
from *HH* and *CH* COSY

```
      46.4
         41.5   34.9
132.2
         50.5        131.9
136.0
            55.1  132.1
         45.5
```

The complete assignment of the C atoms follows from the *CH* correlation (*CH* COSY) and removes any uncertainty concerning the ^{13}C signal assignments in the literature. The *endo*-linkage of the cyclopentene ring to the norbornene residue can be detected from the NOE on the protons at *2.66* and *3.12 ppm*, if the proton 7-H_{syn} at *1.25 ppm* is decoupled. Decoupling of the proton 7-H_{anti} at *1.47 ppm* leads only to NOE enhancement of the bridgehead protons at *2.72* and *2.80 ppm*.

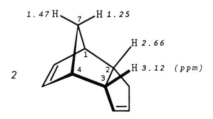

32 *cis*-6-Hydroxy-1-methyl-4-isopropylcyclohexene (carveol)

The correlation signals of the INADEQUATE experiment directly build up the ring skeleton **A** of the compound. Here characteristic ^{13}C shifts (123.1, 137.6; 148.9, 109.1 ppm) establish the existence and position of two double bonds and of one tetrahedral C—O single bond (70.5 ppm). DEPT spectra for the analysis of the *CH* multiplicities become unnecessary, because the INADEQUATE plot itself gives the number of CC bonds that radiate from each C atom.

The *CH* connectivities can be read off from the *CH* COSY plot; thus the complete pattern **B** of all *H* atoms of the molecule is established. At the same time an O*H* group can be identified by the fact that there is no correlation for the broad signal at *4.45 ppm* in the C*H*'COSY plot.

Chemical shifts (ppm, ^{13}C: upright; ^{1}*H*: italics)

A

19.4
137.6
70.5   123.1
38.3   31.5
41.4
148.9
109.1   20.5

B

1.70
4.45 H *4.12*   *5.40*
1.55a * *2.12e* *   *1.77a* * *1.98e* *
2.22
4.70   *1.69*

* *AB systems: a: axial; e : equatorial*

The relative configuration of the O*H* and isopropenyl groups remains to be established. The ^{1}*H* signal at *1.55 ppm*, a C*H*$_2$ proton, splits into a pseudotriplet (*12.4 Hz*) of doublets (*10.1 Hz*). One of the two *12.4 Hz* couplings is the result of the other *geminal* proton of the C*H*$_2$ group; the second of the two *12.4 Hz* couplings and the additional *10.1 Hz* coupling correspond to an *antiperiplanar* relationship of the coupling protons; the *vicinal* coupling partner of the methylene protons is thus located *diaxial* as depicted in the stereoformula **C**, with a *cis* configuration of the O*H* and isopropenyl groups. Hence it must be one of the enantiomers of carveol (**C**) shown in projection **D**.

33 Menthane-3-carboxylic acid (1,3-*cis*-3,4-*trans*-)

The proton *3-H* next to the carboxy group has the largest chemical shift value (*2.30 ppm*). It splits into a pseudotriplet (*13 Hz*) of doublets (*3 Hz*). Since the proton has no *geminal* *H* as coupling partner, only two *antiperiplanar*, i.e. *coaxial*, *H* atoms in the 2- and 4- positions can bring about the two *13 Hz* couplings. The carboxy group is thus in an *equatorial* position (structure **A**). A *syn* proton in the 2-position causes the additional doublet splitting (*3 Hz*).

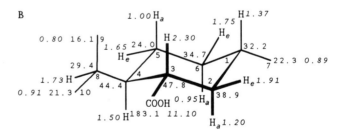

The three coupling partners ($4\text{-}H_a$, $2\text{-}H_a$, $2\text{-}H_e$) of proton *3-H* give three cross signals at *1.20* ($2\text{-}H_a$), *1.50* ($4\text{-}H_a$) and *1.91 ppm* ($2\text{-}H_e$). Its coupling relationships are identifiable from multiplets which are easily analysed; $4\text{-}H_a$ with *1.50 ppm* appears as a pseudotriplet (*13 Hz* with $3\text{-}H_a$ and $5\text{-}H_a$) of pseudotriplets (*2.5 Hz* with $5\text{-}H_e$ and *8-H*). The *axial* proton at C-2 at *1.20 ppm* splits into a pseudoquartet (*13 Hz*). The *geminal* $2\text{-}H_e$ and the two *anti* protons *1-H* and *3-H* are coupling partners.

The assignment of the other multiplets can only be achieved by evaluation of the cross signals in the *HH* COSY plot, and the *CH* COSY plot allows a clear assignment of the *AB* protons which are bonded to methylene C atoms (*1.00* and *1.65* at *24.0*; *0.95* and *1.75* at *34.7*; *1.20* and *1.91* at *38.9 ppm*). It is evident that the *axial* protons always have lower ^1H shift values than their *equatorial* coupling partners on the same C atoms. The assignment of all shifts (stereoformula **B**) and *HH* couplings (Table) can easily be completed. The signals of the diastereotopic methyl groups C-9 and C-10 cannot be assigned.

Chemical shifts (ppm, ^{13}C: upright; ^1H: *italics*)

HH multiplicities, HH couplings (Hz), coupling protons
   (those which are resolved and do not overlap)

$2\text{-}H_a$ *d 13.0* ($2\text{-}H_e$) *d 13.0* (*1-H*)   *d 13.0* (*3-H*)   (*'q'*)
$2\text{-}H_e$ *d 13.0* ($2\text{-}H_a$) *d* 3.0* (*1-H*)   *d* 3.0* (*3-H*)   **('t')*
3-H   *d*13.0* ($2\text{-}H_a$) *d*13.0* (*4-H*) *d 3.0* ($2\text{-}H_e$)   **('t')*
4-H   *d*13.0* (*3-H*) *d*13.0* ($5\text{-}H_a$) *d* 2.5* ($5\text{-}H_e$) *d*2.5* (*8-H*)   **('t','t')*
$7\text{-}H_3$ *d 7.0* (*1-H*)
$9\text{-}H_3$ *d 7.0* (*8-H*)
$10\text{-}H_3$ *d 7.0* (*8-H*)

34 *meso*-α,α,α,α-Tetrakis{2-[(*p*-menth-3-ylcarbonyl)amino]phenyl}porphyrin

By comparison with the data for menthane-3-carboxylic acid (problem 33), the ^{13}C shift values of the menthyl residue A change only slightly when attached to the chiral porphyrin framework. The ^{1}H shift values, however, are noticeably reduced as a result of the shielding effect of the ring current above the plane of the porphyrin ring. The ^{1}H NMR spectrum shows a series of overlapping multiplets between *0.5* and *1.4 ppm* which cannot at first be analysed, so an assignment is possible only with the help of the C*H* COSY plot. With this it is possible to adapt the ^{13}C signal sequence assigned to menthanecarboxylic acid (problem 33) for C-1″ to C-10″. This has been used to generate Table 34.1, where the reference values for menthanecarboxylic acid are in parentheses.

The protons *3″-H (axial)*, *5″-H (axial)*, *7″-H₃* and *8″-H* experience a particularly pronounced shielding. These protons are obviously located well within the range of the shielding ring current above the porphyrin ring plane. This indicates conformation **B** of the molecule, where the isopropyl groups are on the outside and the methyl groups and the *axial* protons *3″-H* and *5″-H* are on the inside.

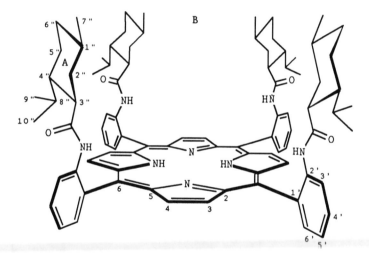

The amide protons, which would otherwise show a considerably larger shift, are also affected by the ring current (*7.15 ppm*). Finally, it is also worth noting that strong shielding of the inner pyrrole N*H* protons (− *2.71 ppm*) is typical for porphyrins.

Table 34.1 Assignment of ^{13}C and ^{1}H chemical shifts (*italics*) of the 'chiral porphyrin lattice'[a]

Porphyrin and phenyl residues			Menthyl residues					
Position	δ_C (ppm)	δ_H (*ppm*)	Position	δ_C ppm	δ_C ppm	δ_H (*ppm*)	δ_H (*ppm*)	$\Delta\delta_H$ (*ppm*)
Pyrrole			1″	28.3	(32.3)	*1.35*	(*1.37*)	−0.02
2,5	132 ± 0.5		2″	38.6	(38.9)	*1.23a*	(*1.20*)	−0.03
3, 4	132 ± 0.5	*8.82–8.89*				*1.35e*	(*1.91*)	−0.56
N*H*		*−2.71*	3″	50.1	(47.8)	*0.85*	(*2.30*)	−1.45 ← [b]
6	115.2		4″	43.9	(44.4)	*1.37*	(*1.50*)	−0.13
Phenyl			5″	23.4	(24.0)	*0.15a*	(*1.00*)	−0.85 ←
1′	130.0					*1.24e*	(*1.65*)	−0.41
2′	138.2		6″	34.0	(34.7)	*0.70a*	(*0.95*)	−0.25
3′	120.9	*8.97*				*1.32e*	(*1.73*)	−0.41
4′	130.1	*7.87*	7″	15.3	(22.3)	*−0.65*	(*0.89*)	−1.54 ←
5′	122.9	*7.43*	8″	31.8	(29.4)	*0.55*	(*1.75*)	−1.20 ←
6′	135.3	*7.72*	9″	[a]21.1	(16.1)	*0.52*	(*0.80*)	−0.28
Amide-N*H*		*7.15*	10″	[a]21.8	(21.3)	*0.55*	(*0.91*)	−0.36
Amide-CO	174.0							

[a]Assignments interchangeable
[b]Arrows indicate significant shieldings due to the porphyrin ring current.

35 *trans*-2-(2-Pyridyl)methylcyclohexanol

The C*H* fragment which is linked to the O*H* group (*5.45 ppm*) can easily be located in the ^{1}H and ^{13}C NMR spectra. The chemical shift values 74.2 ppm for C and *3.16 ppm* for H are read from the C*H* COSY plot. The ^{1}H signal at *3.16 ppm* splits into a triplet (*11.0 Hz*) of doublets (*4.0 Hz*). The fact that an *antiperiplanar* coupling of *11 Hz* appears twice indicates the *diequatorial* configuration (*trans*) of the two substituents on the cyclohexane ring *3*. If the substituents were positioned *equatorial–axial* as in *4* or *5*, then a *synclinal* coupling of *ca 4 Hz* would be observed two or three times.

The pyridine chemical shifts can easily be assigned with the help of the H*H* coupling constants (cf. 2-acetylpyridine, *6*). The ^{13}C chemical shift values of the bonded C atoms can then be read from the C*H* COSY plot. It is more difficult to assign the tetramethylene fragment of the cyclohexane ring because of signal overcrowding. The *geminal AB* systems of the individual CH_2 groups are clearly differentiated in the *CH* COSY plot; the *axial* protons (*1.01–1.22 ppm*) show smaller ^1H shift values than their *equatorial* coupling partners on the same C atom as a result of anisotropic effects; they also show pseudoquartets because of two additional *diaxial* couplings. In the *HH* COSY plot the *HH* connectivities of the *H* atoms attached to C-7—C-2—C-1—C-3 for structure *3* can be identified. Finally, the INADEQUATE plot differentiates between the CH_2 groups in positions 4 and 5 of the cyclohexane ring.

Chemical shifts (ppm, ^{13}C: upright; ^1H: italics)

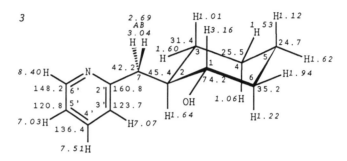

HH multiplicities, HH couplings (Hz), coupling protons (those which are resolved and do not overlap)

1-H	d*11.0	(2-H)	d*11.0	(6-H$_a$)	d4.0	(6-H$_e$)	*('t')
3'-H	d 8.0	(4'-H)					
4'-H	d* 8.0	(3'-H)	d* 8.0	(5'-H)	d2.0	(6'-H)	*('t')
5'-H	d 8.0	(4'-H)	d 5.0	(6'-H)			
6'-H	d 5.0	(5'-H)					

7-H^AH^B form an *AB* system ($^2J_{AB} = 14\ Hz$) of doublets (H^A: $^3J = 5.0$; H^B: $^3J = 4.5\ Hz$) as a result of coupling with *2-H*.

36   2-Hydroxy-3,4,3',4'-tetramethoxydeoxybenzoin, $C_{18}H_{20}O_6$

First, nine double-bond equivalents from the molecular formula, twelve signals in the shift range appropriate for benzenoid C atoms and five multiplets in the shift range appropriate for benzenoid protons, with typical aromatic coupling constants, all indicate a double bond and two benzene rings. Of these two rings, one is 1,2,3,4-tetrasubstituted (*AB* system at *6.68* and *7.87 ppm* and *ortho* coupling of *9 Hz*); the other is 1,2,4-trisubstituted (*ABC* system at *6.79*, *6.87* and *6.97 ppm* with *ortho* and *meta* coupling, *8* and *2 Hz*, respectively). Substituents indicated include:

—in the ^1H NMR spectrum a phenolic *OH* group (*12.34 ppm*),
—in the ^{13}C NMR spectrum a ketonic carbonyl function (203.7 ppm)
—and in both spectra four methoxy groups (*3.68, 3.70, 3.71, 3.87,* and 55.7, 55.7, 56.3 60.1 ppm, respectively), in addition to a methylene unit (*4.26* and 44.3 ppm, respectively).

In order to derive the complete structure, the connectivities found in the *CH* COSY/ *CH* COLOC plots are shown in Table 36.1.

Table 36.1 *CH* connectivities from the *CH* COSY/*CH* COLOC plots

Partial structure	Proton δ_H (*ppm*)	One bond δ_C (ppm)	Two or three bonds δ_C (ppm)	δ_C (ppm)	δ_C (ppm)	δ_C (ppm)
A	7.87	128.0	203.7	158.6	156.5	
B	6.68	104.0	136.0	114.5		
C	6.91	113.6	147.9	121.6		
D	6.87	112.0	148.9	127.5		
E	6.79	121.6	147.9	113.6	44.3	
F	4.26	44.3	203.7	127.5	121.6	113.6
A	3.87	56.3	158.6			
D	3.71	55.7	148.9			
C	3.70	55.7	147.9			
B	3.68	60.1	136.0			

For the 1,2,3,4-tetrasubstituted benzene ring the partial structures **A** and **B** are derived from Table 36.1 from the connectivities of the *AB* protons at *6.68* and *7.87 ppm* and the methoxy protons at *3.68* and *3.87 ppm*. The complete arrangement of the C atoms of the second 1,2,4-trisubstituted benzene ring can be derived from the connectivities **C**, **D** and **E** of the protons of the *ABC* system (*6.79*, *6.87* and *6.97* ppm). From the partially resolved contours of the overlapping correlation signals at *148.9/3.71* and *147.9/3.70* ppm, the methoxy groups at *3.70* and *3.71 ppm* can be identified with the common ^{13}C signal at 55.7 ppm.

Finally, from the partial structures **A** and **F** it can be seen that the two benzene rings are linked to one another by a $-CO-CH_2-$ unit (203.7– 44.3/*4.26 ppm*). Hence it must be 2-hydroxy-3,4-3′,4′-tetramethoxydeoxybenzoin, **G**.

Chemical shifts (ppm, ^{13}C: upright; ^{1}*H: italics*)

G

HH *couplings (Hz)*: $^3J_{5,6} = 9$; $^3J_{5',6'} = 8$; $^4J_{2',6'} = 2$

37 3′,4′,7,8-Tetramethoxyisoflavone, $C_{19}H_{18}O_6$

The molecular formula contains ten double-bond equivalents. In the ^{1}*H* and ^{13}C NMR spectra four methoxy groups can be identified (61.2, 56.7, 57.8 and *3.96, 3.87, 3.78 ppm*, respectively). Of these, two have identical frequencies, as the signal intensity shows (57.8

Table 37.1 Interpretation of the *CH* COSY and *CH* COLOC plots

Partial structure	Proton δ_H (ppm)	One bond δ_C (ppm)	Two or three bonds δ_C (ppm)	δ_C (ppm)	δ_C (ppm)
C	8.48	154.0	175.1	150.1	123.3
A	7.85	121.2	175.1	156.4	150.1
A	7.29	111.2	136.3	118.7	
B	7.19	112.9	148.9	121.5	
B	7.12	121.5	148.9	111.8	
B	6.99	111.8	148.5	124.5	
A	3.96	56.7	156.4		
A	3.87	61.2	136.3		
B	3.78	55.8	148.9	148.5	

and *3.78 ppm*). In the ^1H NMR spectrum an *AB* system (*7.29* and *7.85 ppm*) with *ortho* coupling (*9 Hz*) indicates a 1,2,3,4-tetrasubstituted benzene ring **A**; an additional *ABC* system (*6.99*, *7.12* and *7.19 ppm*) with *ortho* and *meta* coupling (*8.5* and *2 Hz*) belongs to a second 1,2,4-trisubstituted benzene ring **B**. What is more, the ^{13}C NMR spectrum shows a conjugated carbonyl C atom (175.1 ppm) and a considerably deshielded *CH* fragment (154.0 and *8.48 ppm*) with the larger *CH* coupling (198.2 Hz) indicative of an enol ether bond, e.g. in a heterocycle such as furan, *4H*-chromene or chromone.

Knowing the substitution pattern of both benzene rings **A** and **B**, one can deduce the molecular structure from the *CH* connectivities of the *CH* COSY and *CH* COLOC plots. The interpretation of both spectra leads firstly to the correlation Table 37.1.

The benzene rings **A** and **B** derived from the ^1H NMR spectrum can be completed using Table 37.1. The way in which the enol ether is bonded is indicated by the correlation signal of the proton at *8.48 ppm*. The structural fragment **C** results, incorporating the C atom resonating at 123.3 ppm ($^2J_{CH}$), which has not been accommodated in ring **A** or **B** and which is two bonds ($^2J_{CH}$) removed from the enol ether proton.

The combination of the fragments **A**–**C** completes the structure and shows the compound in question to be 3′,4′,7,8-tetramethoxyisoflavone, **D**.

Chemical shifts (ppm, ^{13}C: upright; ^{1}H: italics)

CH multiplicities, *CH* couplings (Hz), coupling *protons*

C-2	D198						
C-3	S	o					
C-4	S	d 6.2	*(2-H)*	d 3.5	*(5-H)*		
C-4a	S	d 8.3	*(6-H)*				
C-5	D163						
C-6	D164						
C-7	S	m					
C-8	S	d 6.0	*(6-H)*	d 3.0	*(2-H)*		
C-8a	S	d*9.2	*(2-H)*	d*9.2	*(5-H)*	*('t')	
C-1'	S	d 7.5	*(5'-H)*				
C-2'	D159	d 7.2	*(6'-H)*				
C-3'	S	m					
C-4'	S	m					
C-5'	D160						
C-6'	D163						
7-OCH$_3$	Q146						
8-OCH$_3$	Q145						
3', 4'-(OCH$_3$)$_2$	Q144						

HH coupling constants (Hz): $^{3}J_{5,6} = 9$; $^{3}J_{5',6'} = 8.5$; $^{4}J_{2',6'} = 2$

38 3',4',6,7-Tetramethoxy-3-phenylcoumarin

Isoflavones *3* that are unsubstituted in the 2-position are characterised in their ^{1}H and ^{13}C NMR spectra by two features:

—a carbonyl-C atom at *ca* 175 ppm (cf. problem 37);
—an enol ether *CH* fragment with high ^{1}H and ^{13}C chemical shift values (*ca 8.5* and 154 ppm) and a remarkably large $^{1}J_{CH}$ coupling constant (*ca* 198 Hz, cf. problem 37).

The NMR spectra of the product do not show these features. The highest ^{13}C shift value is 160.9 ppm and indicates a conjugated carboxy-C atom instead of the keto carbonyl function of an isoflavone (175 ppm). On the other hand, a deshielded *CH* fragment at 138.7 and *7.62 ppm* appears in the ^{13}C NMR spectrum, which belongs to a CC double bond polarised by a $-M$ effect. The two together point to a coumarin *4* with the substitution pattern defined by the reagents.

Table 38.1 Interpretation of the *CH* COSY and *CH* COLOC plots

| | | | C atoms separated by | | | |
| | | | One bond | Two or three bonds | | | |
Partial structure	*Proton* δ_H *(ppm)*	δ_C (ppm)	δ_C (ppm)	δ_C (ppm)	δ_C (ppm)	δ_C (ppm)
C	7.62	138.7	160.9	148.9	127.9	107.8
B	7.22	111.5	149.2	124.3	120.8	
B	7.16	120.8	149.2	124.3	120.8	
A	6.83	107.8	152.3	148.9	146.2	138.7
B	6.81	110.8	148.5	127.6		
A	6.72	99.3	152.3	148.9	146.2	112.2
A	3.86	56.2	152.3			
B	3.85	55.8	148.5			
A	3.84	56.2	146.2			
B	3.82	55.8	149.2			

The correlation signals of the *CH* COSY and the *CH* COLOC plots (shown in the same diagram) confirm the coumarin structure *4*. The carbon and hydrogen chemical shifts and couplings indicated in Table 38.1 characterise rings **A**, **B** and **C**. The connection of the methoxy protons also follows easily from this experiment. The assignment of the methoxy C atoms remains unclear because their correlation signals overlap. Hence the correspondence between the methoxy double signal at 55.8 ppm and the 3′,4′-methoxy signals (55.8 ppm) of 3′,4′,6,7-tetramethoxyisoflavone (problem 37) may be useful until experimental proof of an alternative is found.

Chemical shifts (ppm, ^{13}C: upright; ^{1}H: italics)

4

CH multiplicities, CH couplings (Hz), coupling *protons*

C-2	S	d 8.0	(4-H)				
C-3	S	d*4.0	(2'-H)	d*4.0	(6'-H)	*('t')	
C-4	D160	d 6.0	(5-H)				
C-4a	S	d 6.0	(8-H)				
C-5	D160	d 4.0	(4-H)				
C-6	S	d 7.5	(8-H)	d*3.7	(5-H)	q*3.7	(OCH₃)	*('qui')
C-7	S	d 8.0	(5-H)	d*4.0	(8-H)	q*4.0	(OCH₃)	*('qui)
C-8	D162						
C-8a	S	o					
C-1'	S	d 8.0	(5'-H)	d 4.0	(4-H)	m	
C-2'	D158	d 8.0	(6'-H)				
C-3'	S	d 8.0	(5'-H)	d*4.0	(2'-H)	q*4.0	(OCH₃)	*('qui')
C-4'	S	o					
C-5'	D160						
C-6'	D163	d 8.0	(2'-H)	d 1.0	(5'-H)		
3',4'-(OCH₃)₂	Q 145						
6,7-(OCH₃)₂	Q145						

HH coupling constants (Hz): $^{3}J_{5',6'} = 8$; $^{4}J_{2',6'} = 2$

39 Aflatoxin B₁

The keto-carbonyl ^{13}C signals at 200.9 ppm would only fit the aflatoxins B₁ and M₁. In the ^{13}C NMR spectrum an enol ether-CH fragment can also be recognised from the chemical shift value of 145.8 ppm and the typical coupling constant $J_{CH} = 196$ Hz; the proton involved appears at *6.72 ppm*, as the CH COSY plot shows. The ^{1}H triplet which belongs to it overlaps with a singlet, identified by the considerable increase in intensity of the central component. The coupling constant of the triplet *2.5 Hz* is repeated at *5.39* and *4.24 ppm*. Judging from the CH COSY plot, the proton at *5.39 ppm* is linked to the C atom at 102.5 ppm (Table 39.1); likewise, on the basis of its shift value it belongs to the β-C atom of an enol ether fragment, shielded by the +M effect of the enol ether O atom. The other coupling partner, the allylic proton at *4.24 ppm*, is linked to the C atom at 47.1 ppm, as can be seen from the CH COSY plot (Table 39.1). It appears as a doublet (*7 Hz*) of pseudotriplets (*2.5 Hz*). The larger coupling constant (*7 Hz*) reoccurs in the doublet at *6.92 ppm*. According to the CH COSY plot (Table 39.1), the C atom at 113.5 ppm is bonded to this proton. Hence the evidence tends towards partial structure

Table 39.1 Interpretation of the *CH* COSY/*CH* COLOC plots

		One bond	Two or three bonds			
Partial structure	Proton δ_H (ppm)	δ_C (ppm)	δ_C (ppm)	δ_C (ppm)	δ_C (ppm)	δ_C (ppm)
C	2.46	34.9	200.9			
C	3.22	28.8	177.4			
B	3.91	57.2	161.4			
A	4.24	47.1				
A	5.39	102.5	145.8			
A	6.72	145.8	113.5	102.5	47.1	
B	6.72	91.4	165.1	161.4	107.2	103.5
A	6.92	113.5				

A, and so away from aflatoxin M_1, in which the allylic proton would be substituted by an *OH* group.

Further interpretation of the *CH* COSY/*CH* COLOC plots allows additional assignments to be made for fragments **B** and **C** of aflatoxin B_1.

Since fragment **A** was clearly assigned with the help of *HH* coupling constants, all of the C atoms not included in **A**, which, according to the *CH* COLOC plot, are two or three bonds apart from the equivalent protons at *6.72 ppm* (Table 39.1), belong to the benzene ring **B**.

The assignment of the quaternary C atoms at 154.3, 152.1 and 116.4 ppm has yet to be established. The signal with the smallest shift (116.4 ppm) is assigned to C-11a because the substituent effects of carboxy groups on α-C atoms are small. Since the signal at 152.1 ppm in the coupled spectrum displays a splitting ($^3J_{CH}$ coupling to 9a-H), it is assigned to C-3c.

Additional evidence for the assignment of the other C atoms is supplied by the *CH* coupling constants in the Table shown.

Chemical shifts (ppm, ^{13}C: upright; ^1H: italics)

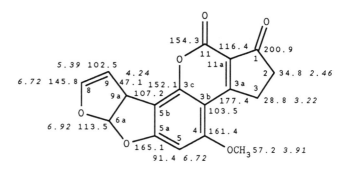

CH multiplicites, *CH* couplings (Hz), coupling *protons*:

C-1	S		t	6.0	(2-H_2)	t	3.0	(3-H_2)				
C-2	T	128.5										
C-3	T	128.5										
C-3a	S		t	5.5	(3-H_2)	t	3.0	(2-H_2)				
C-3b	S		d	5.0	(5-H)							
C-3c	S		d	≤2.5	(9a-H)							
C-4	S		d	*3.5	(5-H)	q*	3.5	(OCH₃)				*('qui)
C-5	D	166.0										
C-5a	S		d	4.5	(9a-H)	d	2.5	(5-H)				
C-5b	S		d	*5.0	(5-H)	d*	5.0	(9-H)				*('t')
C-6a	D	157.5	d	7.5	(9a-H)	d	6.0	(9-H)	d	2.5	(8-H)	
C-8	D	196.0	d	11.0	(9-H)	d*	5.0	(6a-H)	d*	5.0	(9a-H)	*('t')
C-9	D	153.0	d	14.0	(8-H)	d	4.5	(6a-H)	d	2.5	(9a-H)	
C-9a	D	149.0	d	5.5	(8-H)	d	3.5	(6a-H)	d	3.5	(9-H)	*('t')
C-11	S											
C-11a	S		t	3.0	(3-H_2)							
OCH₃	Q	146.5										

HH coupling constants (Hz):
$^3J_{8,9} = {}^3J_{9,9a} = {}^4J_{8,9a} = 2.5; \; {}^3J_{6a,9a} = 7.0$

40   Asperuloside, $C_{18}H_{22}O_{11}$

The molecular formula $C_{18}H_{22}O_{11}$ contains eight double-bond equivalents, i.e. four more than those in the framework *1* known to be present. The ^{13}C NMR spectrum shows two carboxy-CO double bonds (170.2 and 169.8 ppm) and, apart from the enol ether fragment (C-3: 148.9 ppm, $J_{CH} = 194.9$ Hz; C-4; 104.8 ppm, $+M$ effect of the ring O atom), a further CC double bond (C: 142.9 ppm; *CH*: 127.3 ppm); the remaining double-bond equivalent therefore belongs to an additional ring.

Analysis of the *CH* correlation signals (*CH* COSY/*CH* COLOC) for the protons at *7.38* and *5.54 ppm* (Table 40.1) shows this ring to be a five-membered lactone. The *CH* correlation signals with the protons at *4.65 ppm* (*AB* system of methylene protons on C-10) and *2.04 ppm* (methyl group) identify and locate an acetate residue (CO: 170.2 ppm; CH_3: 20.8 ppm) at C-10 (Table 40.1).

Table 40.1 Partial structures from the *CH* COSY and *CH* COLOC plots (the protons are given in *italic* numerals, C atoms separated by a single bond are given in small bold numerals and C atoms separated by two or three bonds are given in small ordinary numerals

CH correlation maxima with the hydrogen atoms at *5.70, 5.54, 4.65, 3.55* and *3.22 ppm* finally establish the position of the additional CC double bond (C-7/C-8, Table 40.1). Hence the structure **A** of the aglycone is now clear.

The iridoid part of the structure C-1—C-9—C-5—C-6—C-7 (**B**) is confirmed by the *HH* COSY plot:

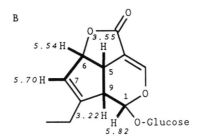

The *¹H* and ¹³C signal assignments of glucopyranoside ring **C** are derived from the *HH* COSY diagram:

$$4.49 \quad 2.98 \quad 3.16 \quad 3.04 \quad 3.18 \quad \begin{matrix} 3.45^A \\ 3.69^B \end{matrix} \quad ppm$$

C   ↑   ↑   ↑   ↑   ↑   ∨

$$98.6 - 73.3 - 76.7 - 70.3 - 77.5 - 61.4 \quad ppm$$

$$C\text{-}1' \quad C\text{-}2' \quad C\text{-}3' \quad C\text{-}4' \quad C\text{-}5' \quad C\text{-}6'$$

As can be seen from a Dreiding model, the five-and six-membered rings of **A** only link *cis* so that a bowl-shaped rigid fused-ring system results. Protons *5-H, 6-H* and *9-H* are in *cis* positions and therefore almost eclipsed. The relative configuration at C-1 and C-9 has yet to be established. Since *1-H* shows only a very small $^3J_{HH}$ coupling (*1.5 Hz*) which is scarcely resolved for the coupling partner *9-H* (*3.22 ppm*), the protons are located in such a way that their CH bonds enclose a dihedral angle of about 120°. The O-glucosyl bond is therefore positioned *synclinal* with respect to *9-H*.

The *antiperiplanar* coupling constant (*8 Hz*) of the protons *1'-H* (*4.49 ppm*) and *2'-H* (*2.98 ppm*) finally shows that a β-glucoside is involved.

The assignment of all of the chemical shift values and coupling constants as derived from the measurements can be checked in structural formula **D**.

Chemical shifts (ppm, ¹³C: upright; *¹H*: italics)

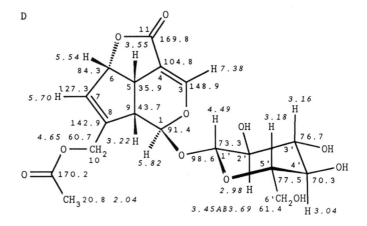

CH multiplicities, *CH* couplings (Hz), coupling *protons*:

C-1	D	179.8	b							
C-3	D	194.9	b							
C-4	S		d*	2.0	(5-H)	d*	2.0	(6-H)	d*	2.0 (9-H) *('q')
C-5	D	149.3	b							
C-6	D	164.9								
C-7	D	169.8	b							
C-8	S		b							
C-9	D	137.6	d	9.5	(7-H, transoid)					
C-10	T	148.0								
C-11	S		d	3.0	(3-H, cisoid)					
C-1'	D	160.3	b							
C-2'	D	138.9	b							
C-3'	D	137.6	b							
C-4'	D	144.2	b							
C-5'	D	140.7	b							
C-6'	T	140.5	b							
Ac-CO	S		q	(CH₃)						
Ac-CH₃	Q	129.7								

HH coupling constants (Hz), where resolved:
$^3J_{1,9} = 1.5$; $^4J_{3,5} = 2.0$; $^3J_{5,6} = 8.0$; $^3J_{5,9} = 8.0$; $^2J_{10-AB} = 14.0$; $^3J_{1',2'} = 8.0$ (anti); $^3J_{2',3'} = 7.5$ (anti); $^3J_{3',4'} = 7.5$ (anti); $^3J_{4',5'} = 8.0$ (anti); $^3J_{5',6'} = 8.0$ (anti); $^3J_{5',6'} = 3.0$ (syn); $^2J_{6'-AB} = 12.5$

The natural product is the asperuloside described in the literature.[39] The assignments for the hydrocarbon pairs C-1/C-1', C-6'/C-10 and C-11/CO (acetyl) have been interchanged. Deviations of ^{13}C chemical shifts (CDCl₃– D₂O[39]) from the values tabulated here [(CD₃)₂SO] are due mainly to solvent effects. Here the difference between the measurements **a** and **d** shows that the use of D₂O exchange to locate the OH protons where the *CH* COSY plot is available is unnecessary since OH signals give no *CH* correlation signal. In this case D₂O exchange helps to simplify the *CH*–OH multiplets and so interpretation of the *HH* COSY plot, which only allows clear assignments when recorded at 600 MHz.

41 9β-Hydroxycostic acid, C₁₅H₂₂O₃

In the ^1H broadband decoupled ^{13}C NMR spectrum, 15 carbon signals can be identified, in agreement with the molecular formula which indicates a sesquiterpene. The DEPT experiments show that the compound contains four quaternary C atoms, three *CH* units, seven *CH₂* units and a *CH₃* group (Table 41.1); this affords the *CH* partial formula C₁₅H₂₀. Consequently, two H atoms are not linked to carbon. Since the molecular formula contains oxygen as the only heteroatom, these two H atoms belong to OH groups (alcohol, carboxylic acid). The ^{13}C NMR spectrum shows a carboxy C atom (170.4 ppm). In the solvent (CD₃OD) the carboxylic proton is not observed because of deuterium exchange. According to *CH* COSY and DEPT, the second OH group belongs to a secondary alcohol (CHOH) with the shifts 80.0 and *3.42 ppm* (Table 41.1).

In the alkene shift range, two methylene groups are found, whose *CH* connectivities are read off from the *CH* COSY plot (Table 41.1, =CH₂: 123.4/5.53 AB 6.18 and =CH₂: 106.9/4.47 AB 4.65). The quaternary alkene C atoms to which they are bonded appear in the ^{13}C NMR spectrum at 146.9 and 151.1 ppm (Table 41.1). Because of the significant difference in the chemical shift values, one of the two CC double bonds (123.4 ppm) must be more strongly polarised than the other (106.9 ppm), which suggests

Table 41.1   Intepretation of the *CH* COSY plot (*CH* fragments)

δ_C (ppm)	CH_n	δ_H (ppm)
170.4	COO	
151.0	C	
146.9	C	
123.4	CH_2	*5.53 AB 6.18*
106.9	CH_2	*4.47 AB 4.65*
80.0	CH	*3.42*
49.8	CH	*1.88*
42.3	C	
38.9	CH_2	*1.23 AB 1.97*
38.5	CH	*2.60*
37.8	CH_2	*2.05 AB 2.32*
36.5	CH_2	*1.53 AB 1.79*
30.8	CH_2	*1.33 AB 1.60*
24.5	CH_2	*1.55 AB 1.68*
11.2	CH_3	*0.75*
CH partial formula	$C_{15}H_{20}$	

that it is linked to the carboxy group ($-M$ effect). The carboxy function and the two
C=CH$_2$ double bonds together give three double-bond equivalents. In all, however, the
molecular formula contains five double-bond equivalents; the additional two evidently
correspond to two separate or fused rings.

Two structural fragments **A** and **B** can be deduced from the *HH* COSY plot; they
include the *AB* systems of *geminal* protons identified from the *CH* COSY diagram (Table
41.1). Fragments **A** and **B** can be completed with the help of the *CH* data in Table 41.1.

The way in which **A** and **B** are linked can be deduced from the *CH* COLOC plot.
There it is found that the C atoms at 80.0 (CH), 49.8 (CH), 42.3 (C) and 38.9 ppm (CH$_2$)
are separated by two or three bonds from the methyl protons at *0.75 ppm* and thus
structural fragment **C** can be derived.

$$2.05^A \quad 1.55^A \quad 1.23^A \qquad\qquad 1.53^A \qquad\qquad 1.33^A$$
$$2.32^B \quad 1.68^B \quad 1.97^B \qquad 3.42 \qquad 1.79^B \quad 2.60 \qquad 1.60^B \quad 1.88 \;\; ppm$$
$$| \qquad\quad | \qquad\quad | \qquad\qquad | \qquad\quad | \qquad\quad | \qquad\quad | \quad\cdot\quad |$$
$$37.8 - 24.5 - 38.9 \qquad 80.0 - 36.5 - 38.5 - 30.8 - 38.9 \;\; ppm$$
$$\mathbf{A} \qquad\qquad\qquad\qquad\qquad \mathbf{B}$$

In a similar way, the linking of the carboxy function with a CC double bond follows
from the correlation of the carboxy resonance (170.4 ppm) with the alkene protons at
5.53 and *6.18 ppm*; the latter give correlation signals with the C atom at 38.5 ppm, as do
the protons at *1.33* and *1.53 ppm*, so that taking into account the molecular unit **A** which
is already known, an additional substructure **D** is established.

Table 41.2 Assembly of the partial structures **A–E** to form the decalin framework **F** of the sesquiterpene

The position of the second CC double bond in the structural fragment **E** follows finally from the correlation of the ^{13}C signals at 37.8 and 49.8 ppm with the ^1H signals at *4.47* and *4.65 ppm*. Note that *trans* protons generate larger cross-sectional areas than *cis* protons as a result of larger scalar couplings.

Table 41.2 combines partial structures **A, B, C, D** and **E** into the decalin framework **F**.

The relative configurations of the protons can be derived from an analysis of all the *HH* coupling constants in the expanded ^1H multiplets. The *trans*-decalin link is deduced from the *antiperiplanar* coupling (*12.5 Hz*) of the protons at *1.33* and *1.88 ppm*. The *equatorial* configuration of the O*H* group is derived from the doublet splitting of the proton at *3.42 ppm* with *12.5* (*anti*) and *4.5 Hz* (*syn*). In a corresponding manner, the proton at *2.60 ppm* shows a pseudotriplet (*12.5 Hz*, two *anti* protons) of pseudotriplets (*4.0 Hz*, two *syn* protons), whereby the *equatorial* configuration of the 1-carboxyethenyl group is established. Assignment of all *HH* couplings, which can be checked in Table 41.3, provides the relative configuration **G** of all of the ring protons in the *trans*-decalin.

The stereoformula **G** is the result; its mirror image would also be consistent with the NMR data. Formula **G** shows the stronger shielding of the *axial* protons compared with their *equatorial* coupling partners on the same C atom and combines the assignments of

Table 41.3 Relative configurations of the protons between *1.23* and *3.42 ppm* from the *HH* coupling constants of the expanded proton multiplets. Chemical shift values (*ppm*) are given as large numerals and coupling constants (*Hz*) are as small numerals

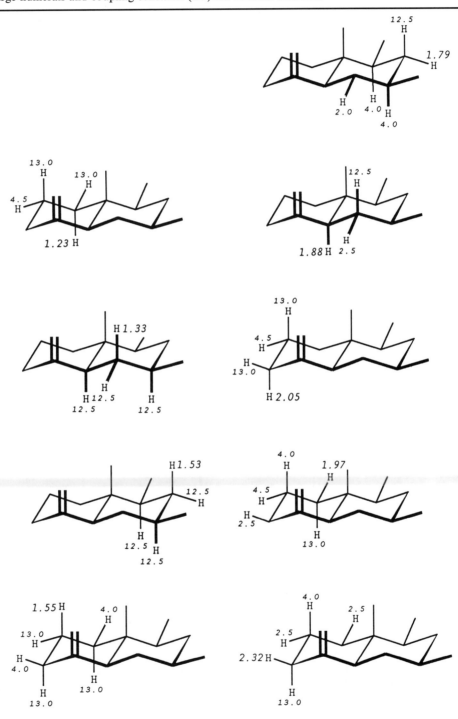

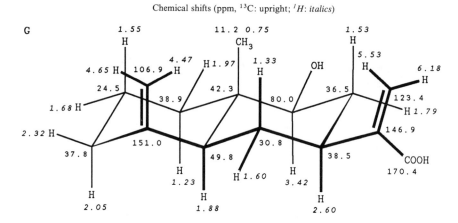

all ^{13}C and ^1H shifts given in Table 41.1. The result is the known compound 9β-hydroxycostic acid.[40]

Chemical shifts (ppm, ^{13}C: upright; ^1H: *italics*)

42   14-(Umbelliferon-7-*O*-yl)driman-3α,8α-diol

The given structure **A** is confirmed by interpretation of the C*H* COSY and C*H* COLOC diagrams. All of the essential bonds of the decalin structure are derived from the correlation signals of the methyl protons. In this, the DEPT spectra differentiate between the tetrahedral C atoms which are bonded to oxygen (75.5 ppm; C*H*—O; 72.5 ppm: C—O; 66.6 ppm: C*H*$_2$—O). The methyl protons at *1.19 ppm*, for example, give correlation maxima with the C atoms at 72.5 ($^2J_{CH}$), 59.4 ($^3J_{CH}$) and 44.1 ppm ($^3J_{CH}$). A corresponding interpretation of the other methyl C*H* correlations ($^3J_{CH}$ relationships) gives the connectivities which are indicated in bold in structure **A**. The assignment of CH_2 groups in positions 2 and 6 remains to be established; this can be done by taking into account the deshielding α-and β-effects and the shielding γ-effects (as sketched in formulae **B** and **C**).

The assignment of the umbelliferone residue in **A** likewise follows from interpretation of the J_{CH} and $^{2,3}J_{CH}$ relationships in the CH COSY and CH COLOC plots following Table 42.1. The ^{13}C signals at 112.9 and 113.1 ppm can be distinguished with the help of the coupled ^{13}C NMR spectrum: 112.9 ppm (C-3′) shows no $^3J_{CH}$ coupling, whereas 113.1 ppm (C-6′) shows a $^3J_{CH}$ coupling of 6 Hz to the proton 8′-H.

Table 42.1   Interpretation of the CH COSY and CH COLOC plots

| | | | C atoms separated by | | | |
| | One bond | | Two or three bonds | | | |
Protons (ppm)	$CH_n{}^a$	ppm	ppm	ppm	ppm	ppm
7.59	CH	143.5	161.3	155.7	128.7	
7.30	CH	128.7	161.8	155.7	143.5	
6.82	CH	101.6	161.8	155.7		
6.80	CH	113.1	161.8	155.7		
6.19	CH	112.9	161.3			
4.13 AB 4.37	CH_2	66.6				
3.39	CH	75.5	32.7			
1.53 AB 1.90	CH_2	44.1^b				
1.53 AB 1.90	CH_2	25.1^b				
1.84	CH	59.4				
1.39 AB 1.65	CH_2	32.7				
1.30 AB 1.55	CH_2	20.0				
1.49	CH	48.8				
1.19	CH_3	24.6	72.5	59.4	44.1	
0.96	CH_3	28.4	75.5	48.4	37.4	22.1
0.90	CH_3	16.0	59.4	48.4	37.9	32.7
0.80	CH_3	22.1	75.5	48.4	37.9	28.4

^{a} CH multiplicities from the DEPT ^{13}C NMR spectra.
^{b} AB systems of the protons attached to these C atoms overlap.

Because of signal overcrowding in the aliphatic range between *1.3* and *2.0 ppm*, the *HH* coupling constants cannot be analysed accurately. Only the deshielded *3-H* at *3.39 ppm* shows a clearly recognisable triplet fine structure. The coupling constant of *2.9 Hz* indicates a dihedral angle of 60° with the protons *2-H^A* and *2-H^B*; thus, *3-H* is *equatorial*. If it were *axial* then a double doublet with one larger coupling constant (*ca 10 Hz* for a dihedral angle of 180°) and one smaller coupling constant (*3 Hz*) would be observed.

Chemical shifts (ppm, ^{13}C: upright; ^{1}H: italics)

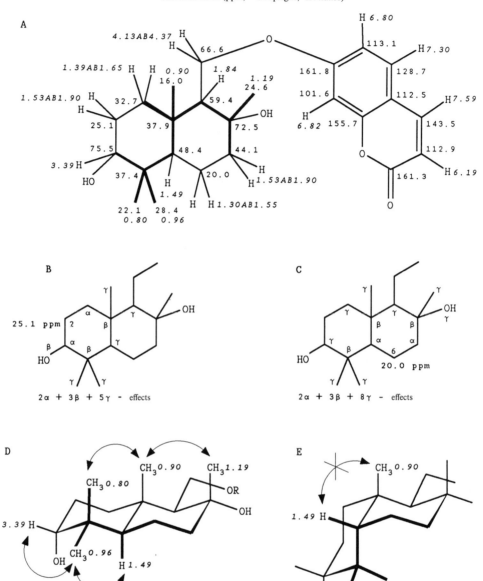

The NOE difference spectra provide more detailed information regarding the relative configuration of the decalin. First, the *trans* decalin link can be recognised from the significant NOE of the methyl-^{1}H signals at *0.80* and *0.90 ppm*, which reveals their *coaxial* relationship as depicted in **D**. For *cis* bonding of the cyclohexane rings an NOE between the methyl protons at *0.90 ppm* and the *cis* bridgehead proton *5-H* (*1.49 ppm*) would be observed, as **E** shows for comparison. An NOE between the methyl protons at *0.90* and *1.19 ppm* proves their *coaxial* relationship, so the 8-O*H* group is *equatorial*.

¹H chemical shifts (ppm)

F

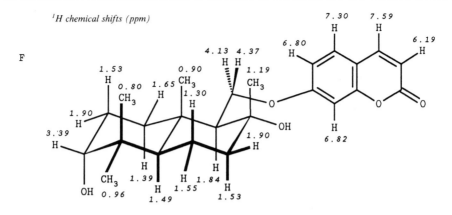

¹³C chemical shifts (ppm)

G

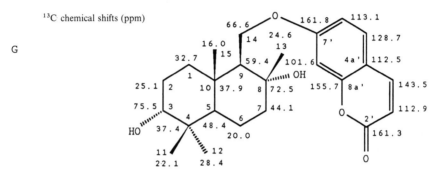

CH multiplicities, *CH* couplings (Hz), coupling *protons*:

C-1	T	126.2
C-2	T	125.7
C-3	D	146.2
C-4	S	
C-5	D	122.0
C-6	T	124.7
C-7	T	125.0
C-8	S	
C-9	D	124.7
C-10	S	
C-11	Q	125.2
C-12	Q	125.7
C-13	Q	125.0
C-14	'T'	143.6
C-15	Q	123.1

C-2'	S									
C-3'	D	173.1								
C-4'	D	163.1	d	5.2	(5'-H)					
C-4a'	S		m							
C-5'	D	162.0	d	3.7	(4'-H)					
C-6'	D	163.1	d	5.2	(8'-H)					
C-7'	S		d	10.0	(5'-H)					
C-8'	D	163.6	d	4.7	(6'-H)					
C-8a'	S		d	10.0	(5'-H)	d*	5.8	(4'-H)	d* 5.8 (8'-H)	*('t')

HH coupling constants (Hz):
$^{3}J_{2A,3} = {}^{3}J_{2B,3} = 2.9$; $^{2}J_{14A,14B} = 9.9$; $^{3}J_{9,14A} = 5.6$; $^{3}J_{9,14B} = 4.0$; $^{3}J_{3',4'} = 9.5$; $^{3}J_{5',6'} = 8.6$;
$^{4}J_{6',8'} = 2.5$

Further effects confirm what has already been established (*5-H* at *1.49 ppm cis* to the methyl protons at *0.96 ppm*; *3-H* at *3.39 ppm syn* to the *geminal* methyl groups at *0.80* and *0.96 ppm*; *9-CH^AH^B* with *H^A* at *4.13* and *H^B* at *4.37 ppm* in spatial proximity to the umbelliferone protons *6'-H* and *8'-H* at *6.80* and *6.82 ppm*). The natural product is therefore 14-(umbelliferon-7-*O*-yl)-driman-3α,8α-diol, **D**, or its enantiomer.

Stereoformula **F** (with *¹H* chemical shifts) and stereoprojection **G** (with ¹³C chemical shifts) summarise all assignments, whereby *equatorial* protons exhibit the larger *¹H* shifts according to their doublet structure which can be detected in the *CH* COSY plot; *equatorial* protons, in contrast to their coupling partners on the same C atom, show only *geminal* couplings, and no additional comparable *antiperiplanar* couplings. The NOE difference spectra also differentiate between the O-*CH₂* protons (*4.13* close to the methyl group at *0.90 ppm*, *4.37* close to the methyl group at *1.19 ppm* as shown in **F**).

43   3,4,5-Trimethyl-5,6-dihydronaphtho[2,3-*b*]furan

The molecular formula $C_{15}H_{16}O$, which indicates a sesquiterpene, contains eight double bond equivalents; in the sp² ¹³C chemical shift range (107.5–154.4 ppm) ten signals appear which fit these equivalents. Since no carboxy or carbonyl signals can be found, the compound contains five CC double bonds. Three additional double bond equivalents then show the system to be tricyclic.

In the ¹³C NMR spectrum the large *CH* coupling constant (197.0 Hz) of the *CH* signal at 141.7 ppm indicates an enol ether unit (=*CH*—O—), as occurs in pyran or furan rings. The long-range quartet splitting ($^3J_{CH}$ = 5.9 Hz) of this signal locates a *CH₃* group in the α-position. This structural element **A** occurs in furanosesquiterpenes, the furano-eremophilanes.

Table 43.1 Interpretation of the *CH* COSY and *CH* COLOC plots

Partial structure	Protons	One bond	Two or three bonds					
	ppm	ppm	ppm	ppm	ppm	ppm	ppm	ppm
B	*1.16*	19.6	27.5	31.1	133.2			
C	*2.30 AB 2.63*	31.1	19.6	27.5	125.5	128.2	133.2	
D	*2.44*	11.4	116.5	126.6	141.7			
E	*2.63*	14.1	107.5	125.3	126.6	127.9	130.0	133.2
F	*3.36*	27.5	19.6	31.1	125.3	127.9	130.0	133.2
G	*5.94*	125.3	27.5	130.0				
H	*6.54*	128.2	31.1	107.5	130.0	133.2		
I	*7.05*	107.5	126.6	128.2	133.2	154.4		
J	*7.33*	141.7	116.5	126.6	154.4			

Starting from the five CC double bonds, three rings and a 3-methylfuran structural fragment, analysis of the *CH* COSY and *CH* COLOC diagrams leads to Table 43.1 and the identification of fragments **B–J**.

CH coupling relationships over two and three bonds (very rarely more) cannot always be readily identified. However, progress can be made with the help of the *CH* fragments which have been identified from the *CH* COSY plot, and by comparing the structural fragments **B–J** with one another. One example would be the assignment of the quaternary C atoms at 130.0 and 133.2 ppm in the fragments **G** and **H**: the weak correlation signals with the proton at *6.54 ppm* may originate from *CH* couplings over two or three bonds; the correlation signal *5.94/130.0* ppm clarifies this; the alkene proton obviously only shows *CH* relationships over three bonds, namely to the C atoms at 130.0 and 27.5 ppm.

The structural fragments **B–J** converge to 3,4,5-trimethyl-5,6-dihydronaphtho[2,3-*b*]furan, **K**. Whether this is the 5(*S*)-or 5(*R*)-enantiomer (as shown) cannot be decided conclusively from the NMR measurements. It is clear, however, that the 5-*CH* proton at

3.36 ppm is split into a pseudoquintet with *7.1 Hz*; this is only possible if one of the 6-*CH₂* protons (at *2.63 ppm*) forms a dihedral angle of about 90 ° with the 5-*CH* proton so that $^3J_{HH} \approx 0$ Hz.

Chemical shifts (ppm, ^{13}C: upright; ^{1}H: italics)

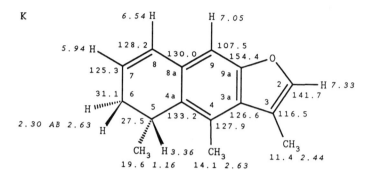

CH multiplicities, *CH* couplings (Hz), coupling *protons*:

C-2	D	197.0	q	5.9	(*3-CH₃*)					
C-3	S		d	12.0	(*2-H*)	d	5.9	(*3-CH₃*)		
C-3a	S		m							
C-4	S		m							
C-4a	S		m							
C-5	D	130.0	m							
C-6	T	128.5	m							
C-7	D	161.5	d*	7.9	(*5-H*)	t*	(*6-H₂*)			*('q')*
C-8	D	157.5	d*	5.9	(*9-H*)	t*	(*6-H₂*)			*('q')*
C-8a	S		m							
C-9	D	161.5	d	3.7	(*8-H*)					
C-9a	S		d	7.9	(*9-H*)					
3-*CH₃*	Q	128.0								
4-*CH₃*	Q	126.0								
5-*CH₃*	Q	126.0	d	5.9	(*6-Hᴬ*)	d*	3.0	(*5-H*)	d*	3.0	(*6-Hᴮ*)	*('t')*

HH coupling constants (Hz):

$^4J_{2,3-Me} = 1.5$; $^3J_{5,6A} = 7.1$; $^3J_{5,5-Me} = 7.1$; $^2J_{6A,6B} = 16.5$; $^3J_{6A,7} = 1.5$; $^3J_{6B,7} = 6.5$;
$^3J_{6B,8} = 3.1$; $^3J_{7,8} = 9.5$

44 6β-Acetoxy-4,4a,5,6,7,8,8a,9-octahydro-3,4aβ,5β-trimethyl-9-oxonaphtho[2,3-*b*]furan-4β-yl-2-methylpropanoic acid ester (Sendarwin)

The high-resolution molecular mass of the compound gives the molecular formula $C_{21}H_{28}O_6$, which corresponds to eight double bond equivalents. The ^1H broadband decoupled ^{13}C NMR spectrum shows a keto carbonyl group (185.2), two carboxy functions (176.4 and 170.4) and four further signals in the sp^2 chemical shift range (146.8, 145.2, 134.3 and 120.9 ppm). These signals identify five double bonds. The three double bond equivalents still missing must then belong to three separate or fused rings. A complete interpretation of the *CH* COSY and DEPT experiments leads to the correlation Table 44.1 and to a *CH* partial molecular formula of $C_{21}H_{28}$, which shows that all 28 *H* atoms of the molecule are bonded to C atoms, and that no *OH* groups are present.

Table 44.1 Interpretation of the CH COSY plot and the DEPT spectra

C (ppm)	CH_n	H (ppm)
185.2	CO^a	
176.4	COO^a	
170.4	COO^a	
146.8	$=C$	
145.2	$=CH$	7.31
134.3	$=C$	
120.9	$=C$	
75.8	$CH-O$	6.29
75.0	$CH-O$	4.88
54.9	CH	2.41
49.5	C	
44.1	CH	1.95
34.5	CH	2.62
29.5	CH_2	1.49 AB 1.95
21.2	CH_3	2.02
18.7	CH_3	1.21
18.6	CH_3	1.21
15.7	CH_2	1.66 AB 2.06
14.6	CH_3	0.92
9.8	CH_3	1.08
8.8	CH_3	1.83
	$C_{21}H_{28}{}^b$	

^{a} Other linkages are eliminated on the basis of the molecular formula.
^{b} CH partial formula obtained by adding CH_n units.

In the HH COSY plot, structural fragment **A** can be identified starting from the signal at *4.88 ppm*. It is evident that two non-equivalent protons overlap at *1.95 ppm*. The CH COSY diagram (expanded section) shows that one of these protons is associated with the CH at 44.1 and the other with the methylene C atom at 29.5 ppm. Altogether the molecule contains two CH_2 groups, identified in the DEPT spectrum, whose methylene AB protons can be clearly analysed in the CH COSY plot and which feature as AB oyotems in the structural fragment **A**.

Resonances in the sp^2 shift region provide further useful information: one at 185.5 ppm indicates a keto carbonyl function in conjugation with a CC double bond; two others at 176.4 and 170.4 ppm belong to carboxylic acid ester groups judging by the molecular formula and since OH groups are not present; four additional signals in the sp^2 shift range (146.8, 145.2, 134.3 and 120.9 ppm) indicate two further CC double bonds, hence the ^1H shift of *7.31 ppm* and the CH coupling constant (202.2 Hz) of the ^{13}C signals at 145.2 ppm identify an enol ether fragment, e.g. of a furan ring with a hydrogen atom attached to the 2-position.

At this stage of the interpretation, the CH correlations across two or three bonds (CH COLOC plot) provide more detailed information. The ^1H shifts given in the CH COLOC diagram, showing correlation maxima with the C atoms at a distance of two to three bonds from a particular proton, lead to the recognition of eight additional structural fragments **B–I** (Table 44.2).

Table 44.2 Partial structures from the CH COLOC plot. Each partial structure **B–I** is deduced from the two- or three-bond couplings J_{CH} for the H atoms of **B–I** (with *italic* ppm values)

Whether the C and H atoms as coupling partners are two or three bonds from one another (^2J or ^3J coupling) is decided by looking at the overall pattern of the correlation signals of a particular C atom with various protons. Thus for methyl protons at *1.08 ppm*, correlation maxima for C atoms are found at 54.9 (CH) and 49.5 ppm (quaternary C). The proton which is linked to the C atom at 54.9 (*2.41 ppm*, cf. CH COSY diagram and Table 44.1) shows a correlation signal with the methyl C at 9.8 ppm, which for its part is linked to the methyl protons at *1.08 ppm*. From this the fragment **J**, which features parts of **C** and **G**, follows directly. The combination of all fragments (following Table 44.2) then gives the furanoeremophilane structure **K**.

The relative configuration of the protons follows from the $^3J_{HH}$ coupling constants, of which it is necessary to concentrate on only two signals (at *4.88* and *2.41 ppm*). The proton at *4.88 ppm* shows a quartet with a small coupling constant (*3 Hz*) which thus has no *antiperiplanar* relationship to one of the *vicinal* protons; it is therefore *equatorial* and this establishes the *axial* position of the acetoxy group. The C*H* proton at *2.41 ppm* shows an *antiperiplanar* coupling (*9 Hz*) and a *synclinal* coupling (*3 Hz*) with the neighbouring methylene protons. From this the relative configuration **L** or its mirror image is derived for the cyclohexane ring.

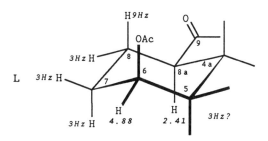

At first the configuration of the methyl groups at C-4a and C-5 remains unclear. The NOE difference spectra, which arise from the decoupling of various *axial* protons, provide the answer. Irradiation at *1.49 ppm* leads to NOE enhancement of the *coaxial* protons (*1.95* and *2.41 ppm*) and of the *cis* protons (*4.88 ppm*). Irradiation at *1.66 ppm* has a strong effect on the methyl group at *1.08 ppm*, and from this the *coaxial* relationship of these protons in the sense of three-dimensional structure **M** is the result. Decoupling at *6.29 ppm* induces strong effects on the *coaxial* protons at *1.95* and *2.41 ppm* and weak effects on the obviously distant methyl groups (*0.92* and *1.08 ppm*); irradiation at *2.41 ppm* has a corresponding effect, producing a very distinct NOE at *6.29 ppm* and a weaker effect at *1.49* and *1.95 ppm*, because in these signals the effects are distributed among multiplet lines. From the *coaxial* relationships thus indicated the structure **M** (or its mirror image with *cis* methyl groups in positions 4a and 5) is deduced.

The stereo projection **N** showing all ^1H and ^{13}C signals summarises all assignments. Again it is evident that *axial* protons (*a*) on the cyclohexane ring are more strongly shielded than their *equatorial* coupling partners (*e*) on the same C atom and that the *diastereotopism* of the isobutanoic acid methyl groups is only resolved in the ^{13}C NMR spectrum.

Chemical shifts (ppm, ^{13}C: upright; ^1H: italics)

45   8α-Acetoxydehydrocostus lactone

In the proton NMR spectrum three pairs of signals appear for alkene CH_2 protons (indicating the three fragments **A**) with *geminal HH* coupling constants ($\leqslant 3\,Hz$) whose assignments can be read off from the *CH* COSY plot:

The *CH* COSY plot in combination with the DEPT subspectrum makes it possible to assign the *AB* systems of all three aliphatic CH_2 protons (structural fragments **B**):

In the structural fragments **A** and **B** all of the *geminal HH* relationships which appear in the *HH* COSY diagram at varying intensities are accounted for. The remaining cross signals concern *HH* connectivities over three bonds (*vicinal* coupling) or four bonds (*long-range* coupling); since the alkene methylene protons do not show *cis* or *trans* coupling, the (*non-geminal*) *HH* connectivities which involve them must belong to couplings over four bonds. Thus all *HH* relationships **C** of the molecule have been established. For ease of reference, the *geminal* relationships are indicated by ||, the *vicinal* by → and $^4J_{HH}$ couplings by |.

The DEPT and *CH* COSY experiments help to complete the proton relationships **C** to the *CH* skeleton **D**; already from **C** it is possible to recognise that the protons *2.44 AB 2.53* and *2.82 ppm*, in common with the alkene methylene group at *5.09 AB 5.28 ppm*, belong to a five-membered ring, whereas the protons at *2.98* and *2.30 AB 2.70 ppm* in common with the alkene methylene group at *4.95 AB 5.05 ppm* belong to a seven-membered ring. From this the guaianolide sesquiterpene skeleton **E** follows, where the links involving oxygen are revealed by the molecular formula and the chemical shifts (*7.42/4.97* and *78.6/4.01 ppm*).

The molecular formula contains eight double bond equivalents, of which **E** has so far accounted for five. The CO single bonds, already identified from the chemical shifts (and inferred from the *CH* balance (DEPT):

$$5C + 5CH + 6CH_2 + CH_3 = C_{17}H_{20}, \text{ cf. } C_{17}H_{20}O_4, \text{ thus no } OH$$

must belong to ester groups since the ^{13}C NMR spectrum identifies two carboxy groups (170.1 ppm and 169.2 ppm). One ester function is then, on the basis of the methyl shifts, an acetate (OCO: 169.2 or 170.1 ppm; CH_3:21.2/*2.15 ppm*); the other

(170.1 or 169.2 ppm) is evidently a lactone ring which with the eighth double bond equivalent completes the constitution **E** to the guaianolide **F**.

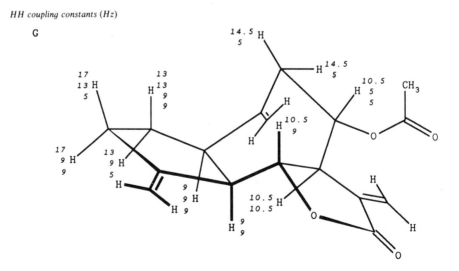

To establish the relative configuration the *HH* coupling constants of suitable multiplets from the *¹H* NMR partial spectrum are best analysed using a Dreiding model. The clearly resolved multiplets at *4.01* and *4.97 ppm* are particularly informative. At *4.01 ppm* a doublet of doublets with *10.5* and *9.0 Hz* indicates an *anti* configuration with respect to the neighbouring protons at *2.82 (9 Hz)* and *3.14 ppm (10.5 Hz)*. At *4.97 ppm* a doublet (*10.5 Hz*) of triplets (*5 Hz*) indicates an *anti* relationship with respect to the neighbouring proton at *3.14 ppm (10.5 Hz)* and *syn* relationships to both *CH₂* protons at *2.30* and *2.70 ppm (5 Hz)*. In the stereostructure **G** drawn from a Dreiding model of the least strained conformer, these relationships can be completed and the principle that the same coupling constant holds for the coupling partner can be verified. The assignment of all chemical shifts is summarised in the stereoprojection **H**, whereby the quaternary C atoms quoted in the literature[43] are assigned by comparison of the data with those of a very similar guaianolide. The chemical shifts of the two carboxy C atoms remain interchangeable (169.2 and 170.1 ppm).

HH coupling constants (Hz)

Chemical shifts (ppm, ^{13}C: upright; ^{1}H: italics)

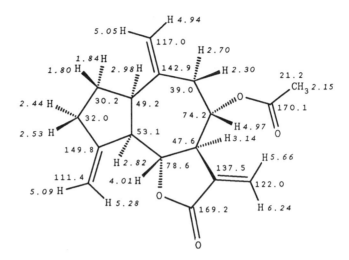

46 Panaxatriol

The sample prepared is not particularly pure, so instead of the 30 signals expected, 33 signals are observed in the ^{1}H broadband decoupled ^{13}C NMR spectrum. Only by pooling information from the DEPT experiment and from the CH COSY plot can a reliable analysis be obtained, as shown in Table 46.1. Here the AB systems of the *geminal* CH_2 protons are assigned.

The three H atoms present in the molecular formula $C_{30}H_{52}O_4$ but missing from the CH balance $C_{30}H_{49}$ (Table 46.1) belong to three hydroxy groups.

Table 46.1   Interpretation of the DEPT spectrum (CH_n) and the CH COSY plot

δ_C(ppm)	CH_n	δ_H(ppm)	δ_C(ppm)	CH_n	δ_H(ppm)
78.5	CH	3.15	36.5	CH_2	1.34 AB 1.50
76.6	C		35.7	CH_2	1.22 AB 1.55
73.2	C		33.1	CH_3	1.20
69.8	CH	3.50	31.1	CH_2	1.03 AB 1.45
68.6	CH	4.08	30.9	CH_3	1.30
61.1	CH	0.87	30.3	CH_2	1.18 AB 1.90
54.7	CH	1.90	27.2	CH_3	1.25
51.1	C		27.1	CH_2	1.55 AB 1.64
49.4	CH	1.40	25.2	CH_2	1.18 AB 1.78
48.7	CH	1.60	19.4	CH_3	1.16
47.0	CH_2	1.53 AB 1.55	17.2	CH_3	0.92
41.0	C		17.2	CH_3	1.04
39.3	C		17.1	CH_3	0.88
39.2	C		16.3	CH_2	1.55 AB 1.77
38.7	CH_2	1.01 AB 1.71	15.5	CH_3	0.97
			CH partial formula $C_{30}H_{49}$		

H

Table 46.2 Interpretation of the NOE difference spectra

Irradiation (ppm)	Significant nuclear Overhauser enhancements (+)										
	0.87	0.92	0.97	1.04	1.30	1.40	1.60	3.15	3.50	4.08	ppm
0.88						+			+		
0.92			+	+						+	
0.97					+					+	
1.04		+					+			+	
1.16							+				
1.30	+		+					+			

Further information is derived from the NOE difference spectra with decoupling of the methyl protons. Table 46.2 summarises the most significant NOE enhancements to complete the picture.

NOE enhancements (Table 46.2) reflect coaxial relationships between

—the CH—O proton at *4.08* and the CH*3* protons at *0.92, 0.97* and *1.04 ppm*,
—the methyl group at *0.88* and the CH protons at *1.40* and *3.50 ppm*,
—the CH proton at *1.60* and the CH*3* protons at *1.04* and *1.16 ppm*,

as well as the *cis* relationship of the CH protons at *0.87* and *3.15* with respect to the methyl group at *1.30 ppm*. From this the panaxatriol structure **A** is derived starting from the basic skeleton of dammarane.

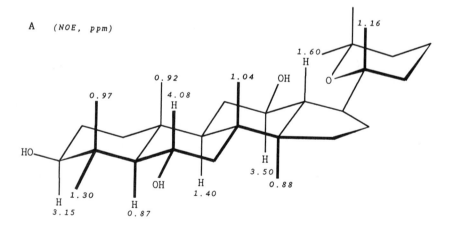

The multiplets and coupling constants of the (*axial*) protons at *3.15, 3.50* and *4.08 ppm* moreover confirm the *equatorial* positions of all three OH groups, as can be seen in formula **B**. Here the couplings from *10.0* to *11.5 Hz*, respectively, identify *vicinal* protons in *diaxial* configurations, whilst values of *4.5* and *5.0 Hz*, respectively, are typical for *axial–equatorial* relationships. As the multiplets show, the protons at *3.50* and *4.08 ppm* couple with two *axial* and one *equatorial* proton (triplet of doublets) respectively, whereas the proton at *3.15 ppm* couples with one *axial* and one *equatorial* proton (doublet of doublets).

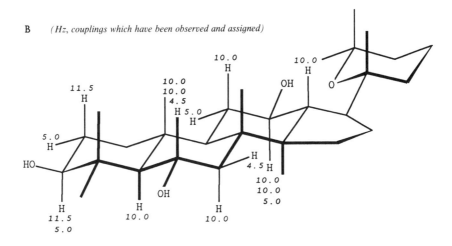

B   *(Hz, couplings which have been observed and assigned)*

Well separated cross signals of the *HH* COSY plot demonstrate

—the *geminal* positions of the methyl groups at *0.97* and *1.30 ppm* and
—the *vicinal* relationship of the protons at *3.15–(1.55 AB 1.64), 1.60–3.50 (1.18 AB 1.90)* and *0.87–4.08 (1.53 AB 1.55)ppm.*

Those C atoms which are bonded to the protons that have already been located can be read from the *CH* COSY plot (Table 46.1) and thus partial structure **C** is the result.

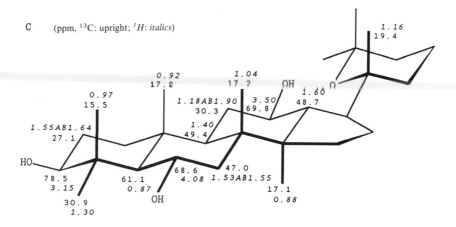

C   (ppm, ¹³C: upright; *¹H*: italics)

The *CH* COLOC diagram shows correlation signals for the methyl protons which are particularly clear. Interpretation of these completes the assignments shown in formula **D** by reference to those *CH* multiplicities which have already been established (Table 46.1).

D   (ppm, ^{13}C: upright; *methyl-^{1}H: italics*)

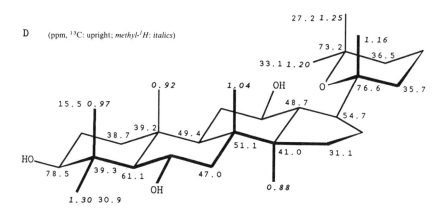

Table 46.3 Interpretation of the C*H* COLOC diagram (methyl connectivities) using the C*H* multiplets derived from Table 46.1

Methyl protons	C atoms separated by two or three bonds							
(*ppm*)	ppm	CH_n	ppm	CH_n	ppm	CH_n	ppm	CH_n
0.88	31.1	CH_2	41.0	C	48.7	CH	51.1	C
0.92	38.7	CH_2	39.2	C	49.4	CH	61.1	CH
0.97	30.9	CH_3	39.3	C	61.1	CH	78.5	CH
1.04	41.0	C	47.0	CH_2	49.4	CH	51.1	C
1.16	35.7	CH_2	54.7	CH	76.6	C		
1.20	27.2	CH_3	36.5	CH_2	73.2	C		
1.25	33.1	CH_3	36.5	CH_2	73.2	C		
1.30	15.5	CH_3	39.3	C	61.1	CH	78.5	CH

E   (ppm, ^{13}C: upright; *^{1}H: italics*)

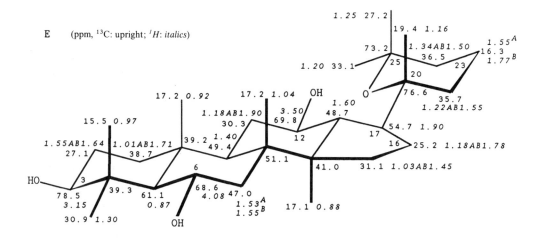

Table 46.3 and formula **D** show that the methyl connectivities of the *CH* COLOC plot are sufficient to indicate essential parts of the triterpene structure.

Differentiation between the methyl groups at 27.2 and 33.1 ppm and between the methylene ring C atoms at 16.3 and 25.2 ppm remains. Here the γ effect on the ^{13}C chemical shift proves its value as a criterion: C-23 is more strongly shielded (16.3 ppm) by the two *axial* methyl groups in (γ) positions 20 and 25 of the tetrahydropyran rings than is C-16 (25.2 ppm). The *axial* CH_3 group at C-25 is correspondingly more strongly shielded (27.2 ppm) than the *equatorial* (33.1 ppm), in accordance with the reverse behaviour of the methyl protons. Thus formula **E** is derived with its complete assignment of all protons and carbon-13 nuclei.

47 4,5-Dimethoxycanthin-6-one (4,5-dimethoxy-6*H*-indolo[3,2,1-*de*]-[1,5]naphthyridin-6-one)

In the NMR spectra two methoxy functions are identified by their typical chemical shift values (partial structures **A**), and their *CH* connectivities are read off from the *CH* COSY plot (Table 47.1).

$$\mathbf{A} \qquad -OCH_3\,4.34 \qquad -OCH_3\,3.94$$
$$61.5 \qquad\qquad 60.7 \quad ppm$$

As expected from the twelve double-bond equivalents implicit in the molecular formula, all other ^{1}H and ^{13}C signals appear in the chemical shift range appropriate for alkenes, aromatics and heteroaromatics ($\delta_H \geqslant 7.4\,ppm$; $\delta_C \geqslant 116\,ppm$).

The *HH* COSY insert indicates two units **B** and **C**:

$$\mathbf{B} \quad 8.36 \quad 7.67 \quad 7.48 \quad 8.21\,ppm \qquad \mathbf{C} \quad 8.73 \quad 8.13\,ppm$$

Their *CH* connectivities can be derived from the *CH* COSY (*CH* COLOC) diagram according to Table 47.1:

```
        8.36    7.67    7.48    8.21 ppm           8.73    8.13 ppm
B        |       |       |       |          C       |       |
      116.0 —130.7 —125.4 —123.4  ppm           145.2 —116.2  ppm
```

The $^3J_{HH}$ (*italic* numbers) and J_{CH} coupling constants (roman numbers) can be deduced and assigned from the ^{1}H NMR and coupled ^{13}C NMR spectra.

Partial structure **B** finally proves to be a 1,2-disubstituted benzene ring because the values of the coupling constants are typical for benzene. Remarkable features of fragment **C** are the large $^1J_{CH}$ coupling constant (181.5 Hz) and the $^3J_{HH}$ coupling (5 Hz), which is considerably smaller than benzenoid *ortho* couplings. These values characterise fragment **C** as a 2,3,4-trisubstituted pyridine ring.

From the CH COLOC plot, the quality of which suffers from a sample concentration that was too low, the two- and three-bond connectivities (Table 47.1) can be read off for fragments **A**, **B** and **C**.

The CH relationships 129.2---*8.21* and 129.2---*8.73* in Table 47.1 are especially valuable for interpretation, because they establish the links between partial structures **B** and **C** and the 4-phenylpyridine skeleton, as it occurs in β-carboline alkaloids **D**.

Table 47.1   Interpretation of the CH COSY/CH COLOC plots

| | | C atoms separated by | | |
| | | One bond | Two or three bonds | |
Partial structure	δ_H *(ppm)*	δ_C *(ppm)*	δ_C *(ppm)*	δ_C *(ppm)*
A	*4.34*	61.5	152.8	
	3.94	60.7	140.9	
B	*8.36*	116.0		
	7.67	130.7	123.4	
	7.48	125.4		
	8.21	123.4	129.2	130.7
C	*8.73*	145.2	133.1	129.2
	8.13	116.2	145.0	128.0

In the β-carboline residue **D** nine double bond equivalents and the partial formula $C_{11}H_6N_2$ are established. Two methoxy groups have already been detected (**A**, $C_2H_6O_2$).

$$C_{11}H_6N_2 + C_2H_6O_2 = C_{13}H_{12}N_2O_2$$

Thus thirteen of the total of sixteen C atoms and all of the H atoms of the total formula are accounted for. However, in comparing this with the molecular formula,

$$C_{16}H_{12}N_2O_3 - C_{13}H_{12}N_2O_2 = C_3O$$

three C atoms in the sp^2 ^{13}C chemical shift range and three double-bond equivalents remain. A carbonyl group appears to fit (carboxamido or carboxy type) in conjugation (157.7 ppm) with a CC double bond (fragment **E**) and a ring which links fragment **E** to the β-carboline heterocyclic **D** with its spare bonds. The positions of the methoxy groups, and the linkages of the C atoms which are bonded to them in the fragment **E** (152.8 and 140.9 ppm), are derived from Table 47.1. To conclude, the substance is 4,5-dimethoxy-canthin-6-one **F**.

All ^{13}C chemical shifts and $^nJ_{CH}$ couplings of the quaternary C atoms can now be assigned in the structure. C-3a, as an example, can be recognised in the ^{13}C NMR spectrum by its typical $^3J_{CH}$ coupling (12.7 Hz) with pyridine proton 2-H.

Chemical shifts (ppm, ^{13}C: upright; ^1H: italics)

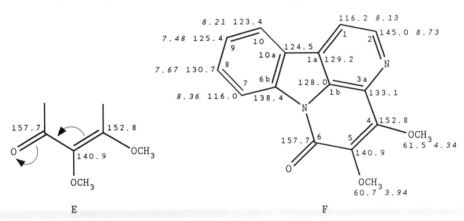

E   F

CH multiplicities, CH couplings (Hz), coupling *protons*:

C-1	168.3	d	7.9	(*2-H*)				
C-1a	S		ü					
C-1b	S		d	7.4 (*1-H*)	d	2.0	(*2-H* or *10-H*)	
C-2	D	181.5	d	3.5 (*1-H*)				
C-3a	S		d	12.7 (*2-H*)				
C-4	S		q	3.5 (*4-OCH₃*)				
C-6	S		s					
C-6b	S		d*	9.3 (*8-H*)	d*	9.3	(*10-H*)	*('t')
C-7	.d	169.5	d	7.5 (*9-H*)				
C-8	D	162.5	d	7.4 (*10-H*)				
C-9	D	163.0	d	6.4 (*7-H*)				
C-10	D	163.9	d	8.7 (*8-H*)				
C-10a	S		d*	7.9 (*7-H*)	d*	7.9	(*9-H*)	*('t')
4-OCH₃	Q	147.0						
5-OCH₃	Q	145.6						

HH coupling constants (Hz):
$^3J_{1,2} = 5.0$; $^3J_{7,8} = {}^3J_{8,9} = {}^3J_{9,10} = 8.0$

48   Cocaine hydrochloride

First the five protons (integral) of the ^1H NMR spectrum (*7.50–7.94 ppm*) in the chemical shift range appropriate for aromatics indicate a monosubstituted benzene ring with typical coupling constants (*8.0 Hz* for *ortho* protons, *1.5 Hz* for *meta* protons.). The chemical shift values especially for the protons which are positioned *ortho* to the substituent (*7.94 ppm*) reflect a $-M$ effect. Using the CH COLOC plot it can be established from the correlation signal *166.4/7.94 ppm* that it is a benzoyl group **A**.

In the HH COSY plot it is possible to take as starting point the peripheral ^1H signal at *5.59 ppm* in order to trace out the connectivities **B** of the aliphatic H atoms:

B   *4.07   2.44   5.59   3.56   4.27   2.51 ppm*

It is then possible to read off from the CH COSY plot those CH links **C** which belong to **B** and to see that between *2.22* and *2.51 ppm* the protons of approximately three methylene groups overlap (integral). Two of these form AB systems in the ^1H NMR spectrum (*2.24 AB 2.44* at 23.7 ppm; *2.22 AB 2.51* at *24.9 ppm*); one pair of the methylene protons approximates the A_2 system (*2.44* at 33.9 ppm) even at 400 MHz.

The ^1H and ^{13}C NMR spectra also indicate an NCH_3 group (*39.6/2.92 ppm*) and an OCH_3 group (*53.4/3.66 ppm*). The CH connectivities **D** of the NCH_3 protons (*2.92 ppm*) across three bonds to the C atoms at 65.3 and 64.5 ppm, derived from the CH COLOC plot, are especially informative, because the combination of **C** and **D** gives the N-methylpiperidine residue **E** with four spare bonds:

The *CH* COLOC diagram also shows

—the linkage **F** of the O*CH₃* proton (*3.66 ppm*) with the carboxy C atom at 174.1 ppm,

$$F \quad \overset{O}{\underset{O - CH_3 \, 3.66}{\underset{174.1}{\Big\Vert}}} \qquad ppm$$

—the connection **G** of the proton at *5.59 ppm* with the same carboxy C atom.

$$G \quad \overset{O}{\underset{\underset{n = 0 \ \ or \ \ 1}{O - CH_3}}{5.59\,H \quad -(C)_n - \overset{174.1}{\Big\Vert}}} \qquad ppm$$

—and the *CH* fragments **H** and **I** involving the protons at *4.07* and *4.27 ppm*; if **B** and **C** are taken into account then the coupling partners (24.9 and *4.07 ppm* and 23.7 and *4.27 ppm*) must be separated by three bonds.

$$4.07\,H - \underset{H}{\overset{}{\underset{\underset{24.9}{CH_2}}{\Big\langle}}} \qquad\qquad \underset{I}{\overset{}{\underset{\underset{23.7}{CH_2}}{\Big\rangle}}} - H\,4.27 \qquad ppm$$

Thus the ecgonin methyl ester fragment **J** can clearly be recognised; only the link to the O atom still remains to be established. The attachment of the O atom is identified by the large chemical shift value (*5.59 ppm*) of the proton on the same carbon atom. The parts **A** and **J** then give the structure **K** of cocaine.

J   K

The fine structure of the *¹H* signal at *5.59 ppm* (*3-H*) reveals the relative configuration of C-2 and C-3. A doublet (*11.5 Hz*) of pseudotriplets (*7.0 Hz*) is observed for an *antiperiplanar* proton (*11.5 Hz*) and two *synclinal* coupling partners (*7.0 Hz*). From that the *cis* configuration of benzoyloxy-and methoxycarbonyl groups is deduced (structure **L**).

The orientation of the N*CH₃* group, whether *syn* or *anti* to the methoxycarbonyl function, is shown by the NOE difference spectrum in tetradeuteriomethanol. If the *N*-methyl proton resonance (*2.92 ppm*) is decoupled an NOE effect is observed for the

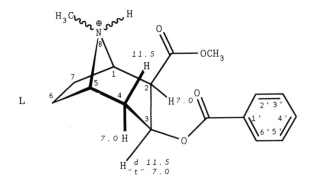

protons at *4.27, 4.07* and between *2.44* and *2.51 ppm* and not merely at *2.44 ppm*. Thus, in tetradeuteriomethanol the *N*-methyl group is positioned *anti* to the methoxy carbonyl group. Hence the assignment of the *endo* and *exo* protons on C-6 and C-7 in the structure **M** of cocaine hydrochloride can also be established.

Chemical shifts (ppm, ^{13}C: upright; ^{1}*II: italics*)

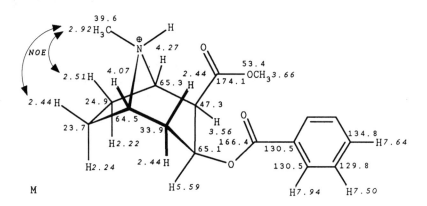

CH multiplicities, *CH* couplings (Hz), coupling *protons*:

C-1	D	155.5									
C-2	D	141.7									
C-3	D	153.6	d*	7.9	*(1-H)*	d*	7.9	*(5-H)*	d*	7.9	*(4-H)*	*('q')*
C-4	T	133.9									
C-5	D	155.5									
C-6	T	135.9									
C-7	T	135.9									
NC*H₃*	Q	143.7									
2-COO	S		b								
OC*H₃*	Q	147.7									
C-1'	S		t	7.9	*(3',5'-H₂)*						
C-2',6'	D	163.4	d*	5.9	*(6'/2'-H)*	d*	5.9	*(4'-H)*			*('t')*
C-3',5'	D	163.4	d	7.9	*(5'/3'-H)*						
C-4'	D	161.4	t	7.9	*(2',6'-H₂)*						
1'-COO	S										

49 Viridifloric acid 7-retronecine ester (heliospathulin)

From the *HH* COSY plot the *HH* relationships **A–D** are read off:

A $4.43 \rightarrow 5.64 \rightarrow \begin{matrix} 1.98^A \\ 2.14^B \end{matrix} \rightarrow \begin{matrix} 2.60^A \\ 3.49^B \end{matrix}$ **B** $\begin{matrix} 3.37^A \\ 4.01^B \end{matrix} \rightarrow 5.60 \rightarrow \begin{matrix} 4.01^A \\ 4.22^B \end{matrix}$

C $3.85 \rightarrow 1.25$ **D** $0.73 \rightarrow 1.93 \rightarrow 0.85$

These can be completed following interpretation of the *CH* COSY diagram (Table 49.1) to give the structural fragments **A–D**.

```
A   4.43 → 5.64 → 1.98^A  → 2.60^A       B   3.37^A → 5.60 →              4.01^A
                  2.14^B    3.49^B           4.01^B                       4.22^B
     |      |       |         |               |         |                   |
 - 76.3 - 76.6 - 34.8   - 53.9  -          - 62.5  - 124.3 = 139.1  -  59.4  -
 -  CH  -  CH  -  CH2   -  CH2  -          -  CH2  -  CH   =   C    -  CH2   -
     |      |                                                  |

C   3.85 → 1.25                           D   0.73 → 1.93 → 0.85    ppm
     |      |                                  |      |      |
    72.5 - 16.6                              17.2 - 31.9 - 15.7     ppm
 -  CH  -  CH3                               CH3  -  CH  -  CH3
     |                                               |
```

Table 49.1 Intepretation of the *CH* COSY and the *CH* COLOC plots and the DEPT spectra

		C atoms separated by			
		One bond		Two or three bonds	
Partial structure	*Protons* δ_H *(ppm)*	(ppm)	δ_C $CH_n)$[a]	δ_C (ppm)	δ_C
A	*5.64*	76.6	*CH*	174.4	76.3
B	*5.60*	124.3	*CH*		
A	*4.43*	76.3	*CH*		
B	*4.01 AB 4.22*	59.4	CH_2	139.2	
C	*3.85*	72.5	*CH*		
B	*3.37 AB 4.01*	62.05	CH_2	139.1	124.3
A	*2.60 AB 3.49*	53.9	CH_2		
A	*1.98 AB 2.14*	34.8	CH_2		
D	*1.93*	31.9	*CH*		
C	*1.25*	16.6	CH_3		
D	*0.85*	15.7	CH_3		
D	*0.73*	17.2	CH_3	83.9	31.9
H atoms bonded to C:			H_{22}	(therefore 3 *OH*)	

[a] CH_n multiplets given by DEPT.

The molecular formula contains four double-bond equivalents, of which the ^{13}C NMR spectrum identifies a carboxy group (174.4 ppm) and a CC double bond (139.1 ppm: C, and 124.3 ppm: CH with H at *5.60 ppm* from CH COSY) on the basis of the three signals in the sp^2 chemical shift range. The two additional double-bond equivalents must therefore belong to two separate or fused rings. Since fragments **A** and **B** terminate in electronegative heteroatoms judging from their ^{13}C (62.5, 53.9 and 76.3 ppm) and ^{1}H chemical shift values, a pyrrolizidine bicyclic system **E** is suggested as the alkaloid skeleton, in line with the chemotaxonomy of *Heliotropium* species, in which fragments **A** and **B** are emphasised with bold lines for clarity.

Correlation signals of the *AB* systems *4.01^{A}4.22^B* and *3.37^{A}4.01^B ppm* with the C atoms of the double bond (at 124.3 and 139.1 ppm, Table 49.1) confirm the structural fragment **B**. A signal relating the proton at *5.64 ppm* (Table 49.1) to the carboxy C atom (at 174.4 ppm) shows that the O*H* group at C-7 has esterified (partial formula **F**) in accord with the higher ^{1}H shift (at *5.64 ppm*) of proton 7-*H* caused by the carboxylate. When the O*H* group at C-7 is unsubstituted as in heliotrin then 7-*H* appears at *4.06 ppm*.[31] On the other hand, the chemical shift values of the *AB* protons at C-9, which are considerably lower than those of heliotrin, indicate that the *9-OH* group is not esterified.

The relative configuration at C-7 and C-8 cannot be established from the *HH* coupling constants; for five-membered rings the relationships between dihedral angles and coupling constants for *cis* and *trans* configurations are not as clear as for six-membered rings. However, NOE difference spectra shed light on the situation: decoupling at *5.64 ppm* (*7-H*) leads to a very distinct NOE at *4.43 ppm* (*8-H*) and vice versa. The protons *7-H* and *8-H* must therefore be positioned *cis*. Decoupling of *7-H* also leads to an NOE on the protons *6-H^A* and *6-H^B*, which indicates the spatial proximity of these three protons. A Dreiding model shows that the envelope conformation of the pyrrolidine ring (C-7 lies out of the plane of C-8, N, C-5 and C-6) in fact places *7-H* between *6-H^A* and *6-H^B* so that the distances to these protons do not differ substantially. The *7-H* signal splits accordingly into a pseudotriplet with *3.5 Hz*; *8-H* and *6-H^A* are coupling partners of *7-4* (dihedral angle *ca* 60°), whilst *6-H^B* and *7-H* have a dihedral angle of 90° so no more couplings are detected.

Finally, fragments **C** and **D** belong to the acidic residues in the alkaloid ester. Taking into account the two O*H* groups (cf. Table 49.1), the C*H* correlation signal of the methyl protons at *0.73 ppm* with the quaternary C atoms at 83.9 ppm links **C** and **D** to the diastereomers viridifloric or trachelanthinic acid, (distinction between the two is discussed in more detail in the literature[31]). The diastereotopism of the isopropyl methyl C atoms is a good criterion for making the distinction. Their chemical shift difference is found to be $\Delta\delta_C = 17.2 - 15.7 = 1.5$ ppm, much closer to the values reported for viridifloric acid ester ($\Delta\delta_C \approx 1.85$ ppm; for trachelanthinic acid ester a value of $\Delta\delta_C \leqslant$ 0.35 ppm would be expected). Thus structure **H** of the pyrrolizidine alkaloid is established. It can be described as viridifloric acid-7-retronecine ester or, because of its plant origin, as heliospathulin.

Chemical shifts (ppm, ^{13}C: upright; ^{1}H: italics)

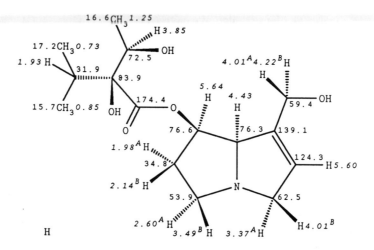

50 *trans*-*N*-Methyl-4-methoxyproline

Rather large *HH* coupling constants in the aliphatics range (*12.5* and *15.0 Hz*) indicate *geminal* methyl protons in rings. In order to establish clearly the relevant *AB* systems, it makes sense first to interpret the *CH* COSY diagram (Table 50.1). From this, the compound contains two methylene groups, **A** and **B**.

$$\mathbf{A} \quad \begin{matrix} 2.34^A \ 2.66^B \\ 38.2 \end{matrix} \qquad \mathbf{B} \quad \begin{matrix} 3.45^A \ 4.02^B \quad ppm \\ 75.4 \qquad ppm \end{matrix}$$

Taking these methylene groups into account, interpretation of the *HH* COSY plot leads directly to the *HH* relationships **C** even if the protons at *2.34* and *4.58 ppm* do not show the expected cross signals because their intensity is spread over the many multiplet lines of these signals.

$$\begin{matrix} & & \mathbf{A} & & \mathbf{B} \\ & & \downarrow & & \downarrow \\ \mathbf{C} & & 2.34^A & & 3.45^A \\ & \to 4.32 \to & 2.66^B & \to 4.58 \to & 4.02^B \to ppm \end{matrix}$$

The *CH* COSY plot completes the *HH* relationships **C** of the *CH* fragment **D**:

$$\begin{matrix} 4.32 & \begin{matrix} 2.66^B \\ 2.34^A \end{matrix} & 4.58 & \begin{matrix} 4.02^B \\ 3.45^A \end{matrix} & ppm \\ | & | & | & | & \\ \mathbf{D} \quad - \ 77.8 \ - & 38.2 & - \ 67.7 \ - & 75.4 & - \quad ppm \\ \\ - \ CH \ - & CH_2 & - \ CH \ - & CH_2 & - \\ | & & | & & \end{matrix}$$

Table 50.1   Intepretation of the *CH* COSY and the *CH* COLOC plots

			C atoms separated by		
		One bond		Two or three bonds	
Partial structure	*Protons* δ_H *(ppm)*	δ_C (ppm)	$CH_n)^a$	δ_C (ppm)	δ_C (ppm)
	4.58	67.7	CH		
	4.32	77.8	CH	170.4	
B	3.45 AB 4.02	75.4	CH_2	67.7	
	3.40	55.0	OCH_3	67.7	
	3.15	49.1	NCH_3	77.8	75.4
A	2.34 AB 2.66	38.2	CH_2	77.8	

[a] CH_n multiplets given by DEPT spectrum.

The typical chemical shift values and CH coupling constants in the one-dimensional NMR spectra reveal three functional groups:

—N-methylamino (—NCH_3, δ_C = 49.1 ppm; Q, J_{CH} = 144 Hz; δ_H = *3.15 ppm*),
—methoxy (—OCH_3, δ_C = 55.0 ppm; Q, J_{CH} = 146 Hz; δ_H = *3.40 ppm*),
—carboxy-/carboxamido-(—COO—, δ_C = 170.6 ppm).

If it were a carboxylic acid, the carboxy proton would not be visible because of deuterium exchange in the solvent tetradeuteriomethanol

$$(—COOH + CD_3OD = —COOD + CD_3OH).$$

In the CH COLOC plot (Table 50.1) the correlation signals of the N—CH_3 protons (*3.15 ppm*) with the terminal C atoms of the CH fragment **D** (75.4 and 77.8 ppm) indicate an N-methylpyrrolidine ring **E**.

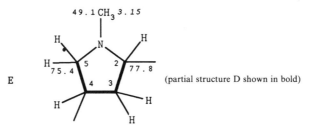

E   (partial structure D shown in bold)

Since the carboxy-C atom in the CH COLOC diagram (Table 50.1) shows no correlation signal with the methoxy protons, it must be a carboxylic acid rather than a methyl ester. In the CH COLOC plot of cocaine (problem 48) there is a cross signal relating the carboxy-C atom with the OCH_3 protons, because cocaine is a methyl ester. Finally, a cross signal relating C-4 (67.7 ppm) of the pyrrolidine ring **E** to the OCH_3 protons (*3.40 ppm*) in the CH COLOC plot locates the methoxy group on this C atom. Hence the structure has been established; it is therefore N-methyl-4-methoxyproline, **F**.

F

The relative configuration is derived from the NOE difference spectra. Significant NOEs are found owing to *cis* relationships within the neighbourhood of non-*geminal* protons:

2.34 ⇆ 4.32; 2.66 ⇆ 4.58; 4.02 ⇆ 4.58; (NCH₃) 3.15 ⇆ 4.02 ppm

From this, the N-methyl and carboxy groups are in *cis* positions whereas the carboxy and methoxy groups are *trans* and so *trans-N*-methyl-4-methoxyproline, **G**, is the structure implied. The NMR measurements do not provide an answer as to which enantiomer it is, 2R,4S as shown or the mirror image 2S,4R.

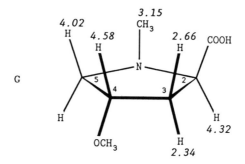

The formulae **H** and **I** summarize the results with the complete assignments of all ^{13}C and ^{1}H chemical shifts (**H**) and the *HH* multiplets and coupling constants (**I**). Here the ^{1}H multiplets which have been interpreted because of their clear fine structure are indicated by the multiplet abbreviation *d* for *doublet*.

Chemical shifts (ppm, ^{13}C: upright; ^{1}H: *italics*)

HH multiplicities and coupling constants (Hz)

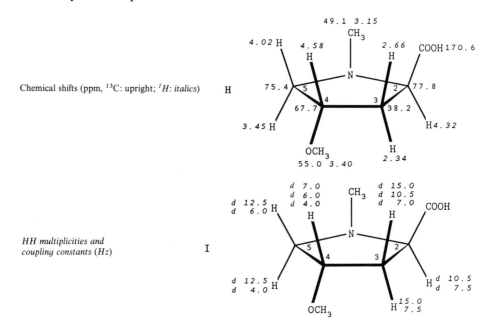

5 REFERENCES

1. M. Hesse, H. Meier and B. Zeeh, *Spektroskopische Methoden in der Organischen Chemie*, 4th edn., Georg Thieme, Stuttgart, 1991.
   see also
   D. H. Williams, I. Fleming, *Spectroscopic Methods in Organic Chemistry*, 7th edn, McGraw-Hill (UK), London, 1989
2. H. Günther, *NMR Spektroskopie, Grundlagen, Konzepte und Anwendungen der Protonen- und Kohlenstoff-13-Kernresonanz-Spektroskopie in der Chemie*, 3rd edn, Georg Thieme, Stuttgart, 1991.
3. H. Friebolin, *One- and Two-dimensional NMR Spectroscopy. An Introduction*, VCH, Weinheim, 1992.
4. G. C. Levy, R. Lichter and G. L. Nelson, *Carbon-13 Nuclear Magnetic Resonance Spectroscopy*, 2nd edn, Wiley-Interscience, New York, 1980.
5. H. O. Kalinowski, S. Berger and S. Braun, *^{13}C NMR Spectroscopy*, J Wiley & Sons, Chichester, 1988.
6. E. Breitmaier and W. Voelter: *Carbon-13 NMR Spectroscopy—High Resolution Methods and Applications in Organic Chemistry and Biochemistry*, 3rd edn, VCH Weinheim, 1990.
7. G. C. Levy and R. L. Lichter, *Nitrogen-15 Nuclear Magnetic Resonance Spectroscopy*, Wiley-Interscience, New York, 1979.
8. E. Breitmaier, Die Stickstoff-15-Kernresonanz— Grenzen und Möglichkeiten, *Pharm. Unserer Zeit*, **12**, 161 (1983).
9. W. von Philipsborn and R. Müller, ^{15}N-NMR-Spektroskopie—neue Methoden und ihre Anwendung, *Angew. Chem.*, **98**, 381 (1986).
10. C. LeCocq and J. Y. Lallemand, *J. Chem. Soc., Chem. Commun.*, 150 (1981).
11. D. W. Brown, T. T. Nakashima and D. I. Rabenstein, *J. Magn. Reson.*, **45**, 302 (1981).
12. A. Bax, *Two-Dimensional Nuclear Magnetic Resonance in Liquids*, Reidel, Dordrecht, 1984.
13. R. R. Ernst, G. Bodenhausen and A. Wokaun, *Principles of Nuclear Magnetic Resonance in One and Two Dimensions*, University Press, Oxford, 1990.
14. D. M. Dodrell, D. T. Pegg and M. R. Bendall, *J. Magn. Reson.*, **48**, 323 (1982); *J. Chem. Phys.* **77**, 2745 (1982).
15. M. R. Bendall, D. M. Dodrell, D. T. Pegg and W. E. Hull, *DEPT, Information brochure with experimental details*, Bruker Analytische Messtechnik, Karlsruhe, 1982.
16. J. L. Marshall, *Carbon–Carbon and Carbon–Proton NMR Couplings: Applications to Organic Stereochemistry and Conformational Analysis*, Verlag Chemie International, Deerfield Beach, FL, 1983.
17. J. K. M. Sanders and B. K. Hunter, *Modern NMR Spectroscopy. A Guide for Chemists*, 2nd edn, Oxford University Press, Oxford, 1993. The authors give a clear introduction to experimental techniques which use the most important one- and two-dimensional NMR methods.
18. W. Aue, E. Bartholdi and R. R. Ernst, *J. Chem. Phys.*, **64**, 2229 (1976).
19. A. Bax and R. Freeman, *J. Magn. Reson.*, **42**, 164 (1981); **44**, 542 (1981).
20. A. Bax, R. Freeman and S. P. Kempsell, *J. Am. Chem. Soc.*, **102**, 4581 (1980).
21. T. H. Mareci and R. Freeman, *J. Magn. Reson.*, **48**, 158 (1982).
22. D. L. Turner, *J. Magn. Reson.*, **49**, 175 (1982).

23. A. Bax and G. Morris, *J. Magn. Reson.*, **42**, 501 (1981).

24. H. Kessler, C. Griesinger, J. Zarbock and H. Loosli, *J. Magn. Reson.*, **57**, 331 (1984).

25. D. Leibfritz, *Chem. Ber.*, **108**, 3014 (1975).

26. J. H. Noggle and R. E. Schirmer, *The Nuclear Overhauser Effect*, Academic Press, London, 1971; D. Neuhaus and M. Williamson, *The Nuclear Overhauser Effect in Structural and Conformational Analysis*, VCH, Weinheim, 1989.

27. M. Kinns and J. K. M. Sanders, *J. Magn. Reson.*, **56**, 518 (1984).

28. G. Bodenhausen and R. R. Ernst, *J. Am. Chem. Soc.*, **104**, 1304 (1982).

29. A. Ejchardt, *Org. Magn. Reson.* **9**, 351 (1977).

30. J. L. C. Wright, A. G. McInnes, S. Shimizu, D. G. Smith, J. A. Walter, D. Idler and W. Khalil, *Can. J. Chem.*, **56**, 1898 (1978).

31. S. Mohanray and W. Herz, *J. Nat. Prod.*, **45**, 328 (1982).

32. W. H. Pirkle and D. J. Hoover, *Top. Stereochem.*, **13**, 263 (1982).

33. G. R. Sullivan, *Top. Stereochem.*, **9**, 111 (1976).

34. M. Gosmann and B. Franck, *Angew. Chem.*, **98**, 1107 (1986); G. Kübel and B. Franck, *Angew. Chem.*, **100**, 1203 (1988).

35. H. Kessler, *Angew. Chem.*, **82**, 237 (1970).

36. J. Sandström, *Dynamic NMR Spectroscopy*, Academic Press, New York, 1982.

37. M. Oki (Ed.), *Applications of Dynamic NMR Spectroscopy to Organic Chemistry*, VCH, Deerfield Beach, FL, 1985.

38. F. McCapra and A. I. Scott, *Tetrahedron Lett.*, 869 (1964).

39. S. Damtoft, S. R. Jensen and B. J. Nielsen, *Phytochemistry*, **20**, 2717 (1981).

40. F. Bohlmann, J. Jakupovic, A. Schuster, R. King and H. Robinson, *Phytochemistry*, **23**, 1445 (1984); S. Sepulveda-Boza and E. Breitmaier, *Chem. Ztg.*, **111**, 187 (1987).

41. F. Graf and M. Alexa, *Planta Med.*, 428 (1985).

42. M. Garrido, S. Sepulveda-Boza, R. Hartmann and E. Breitmaier, *Chem. Ztg.*, **113**, 201 (1989).

43. A. Ortega and E. Maldonado, *Phytochemistry*, **23**, 1507 (1984); R. E. Negrete, N. Backhouse, A. San Martin, B. K. Cassels, R. Hartmann and E. Breitmaier, *Chem. Ztg.*, **112**, 144 (1988).

44. S. Shibata, O. Tanaka, K. Soma, Y. Iida, T. Ando and H. Nakamura, *Tetrahedron Lett.*, 207 (1965); O. Tanaka and S. Yahara, *Phytochemistry*, **17**, 1353 (1978).

45. T. Ohmoto and K. Koike, *Chem. Pharm. Bull.*, **32**, 3579 (1984).

FORMULA INDEX OF SOLUTIONS TO PROBLEMS

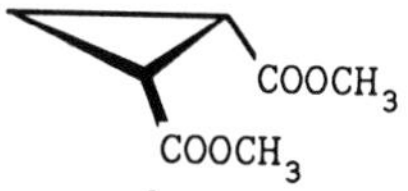

1

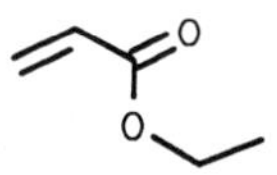

2

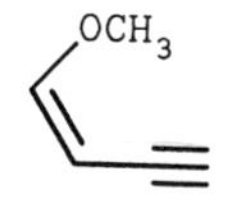

3

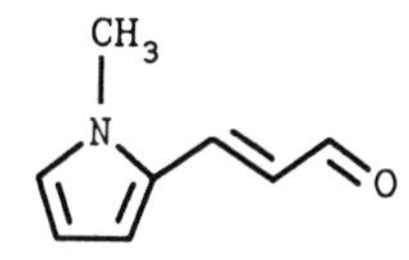

4

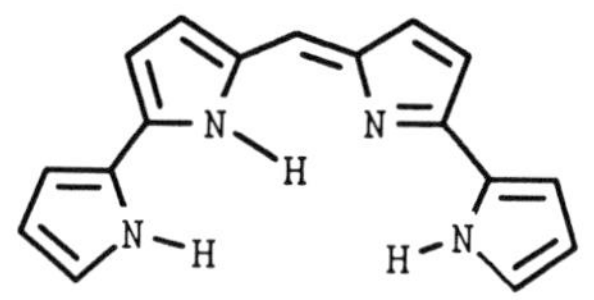

5

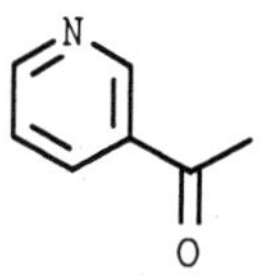

6

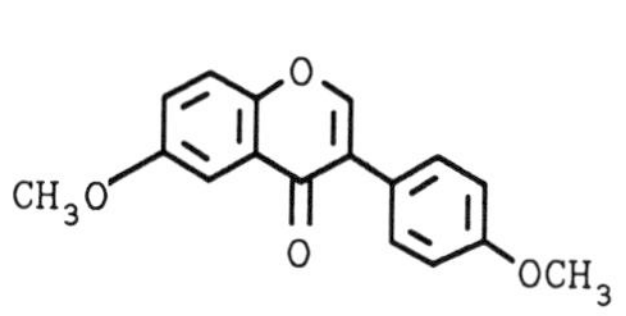

7

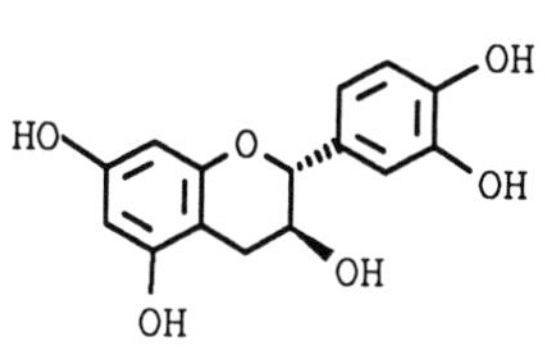

8

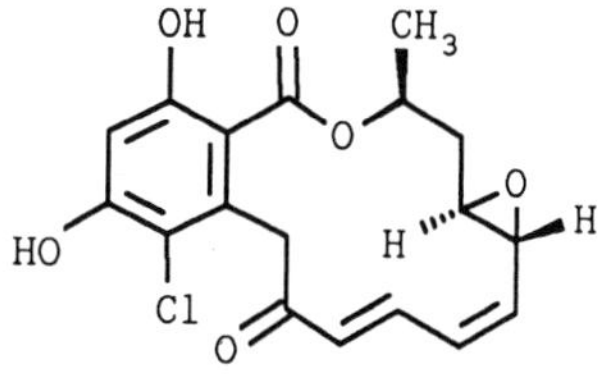

9

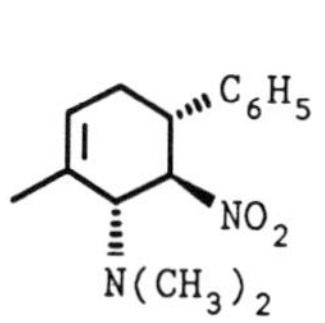

10

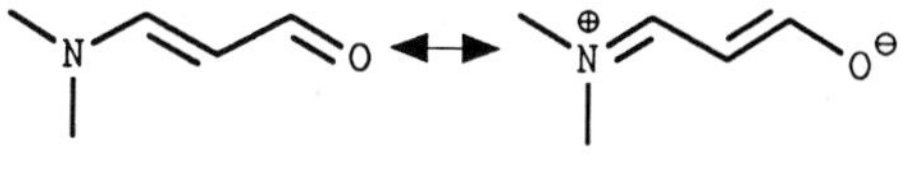

11

$\Delta G = 64.7$ kJ/mol

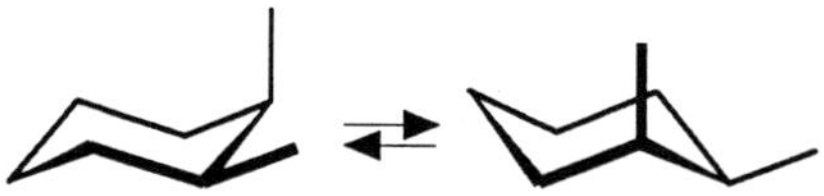

12

$\Delta G = 40.8 \ kJ/mol$

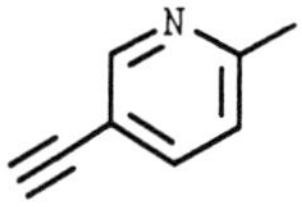

13

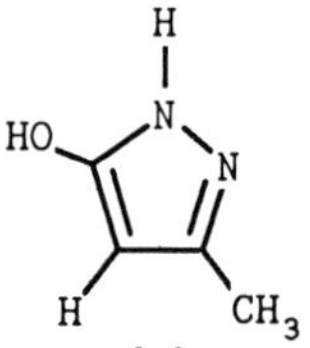

14

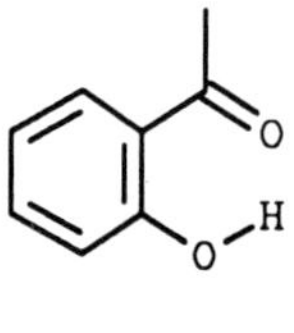

15

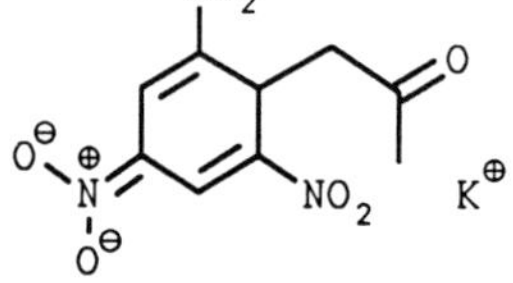

16

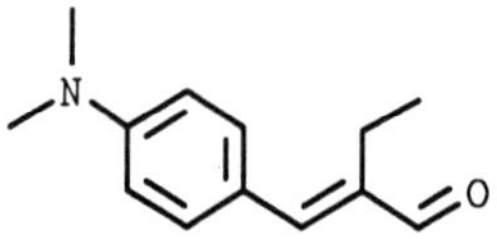

17

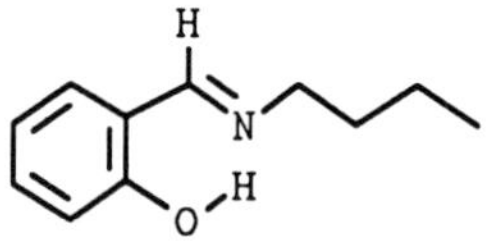

18

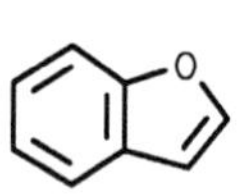

19

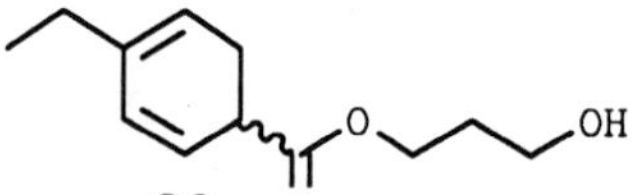

20

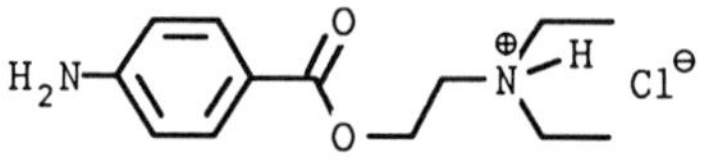

21

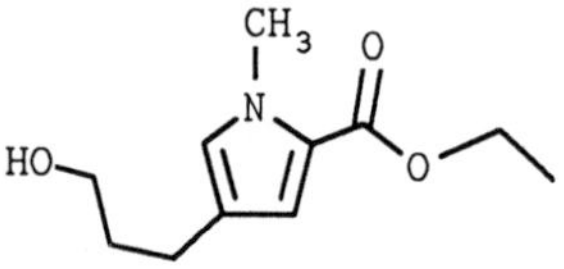

22

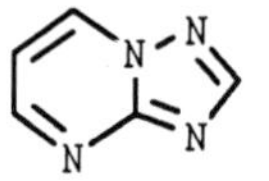

24

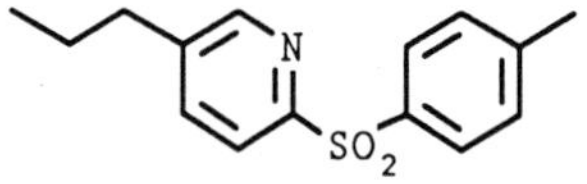

23

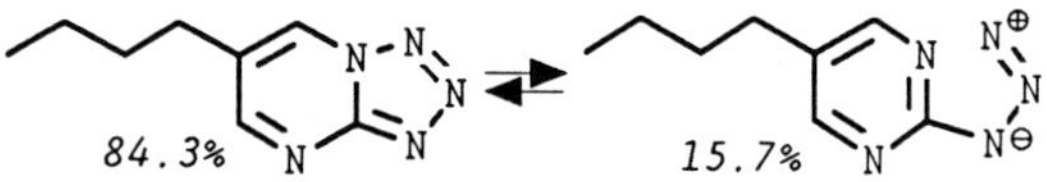

84.3%   15.7%

25

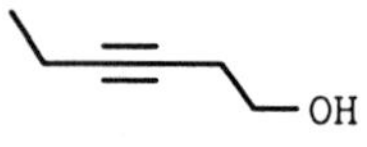

26

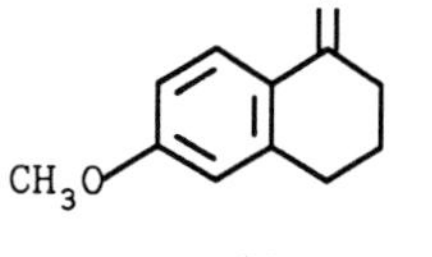

27

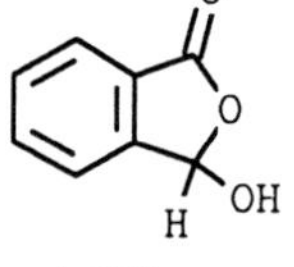

28

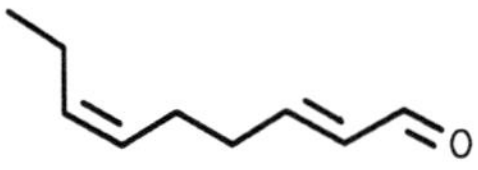

29

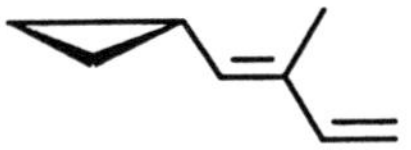

30

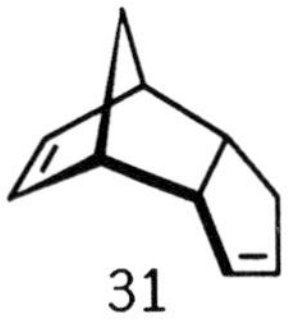

31

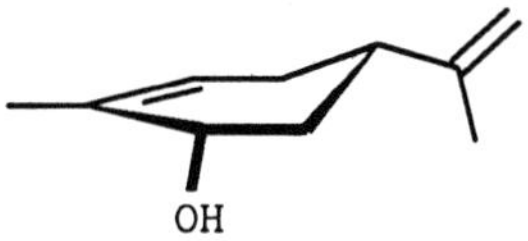

32

33

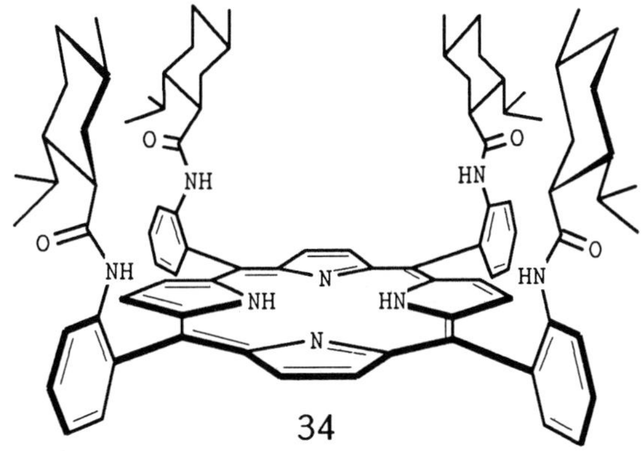

34

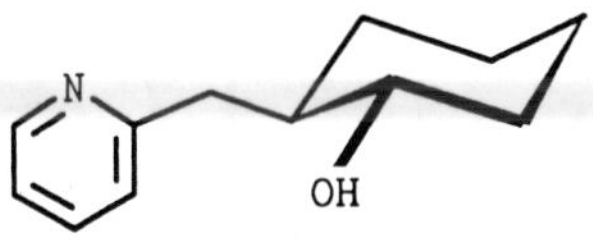

35

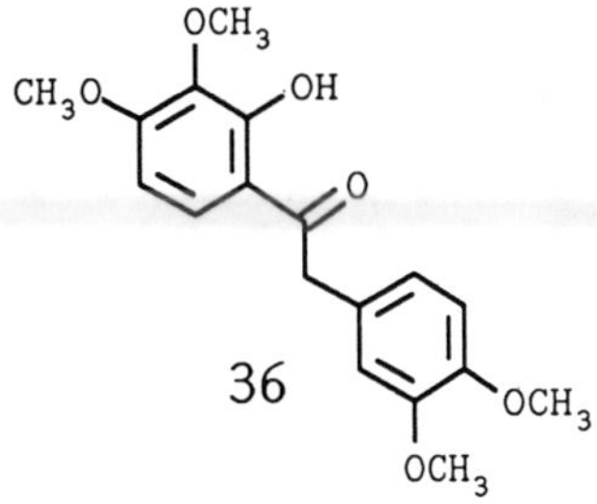

36

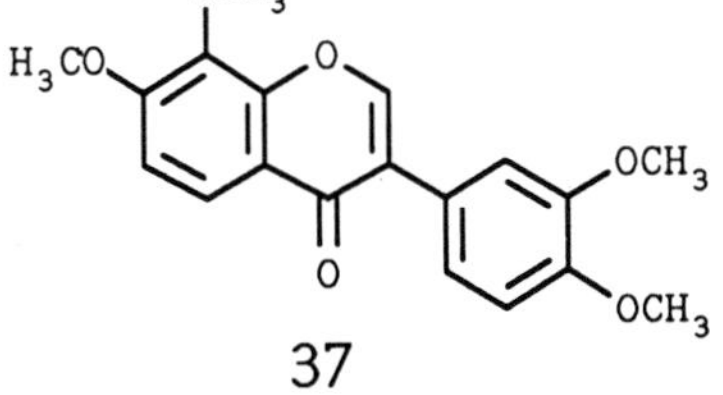

37

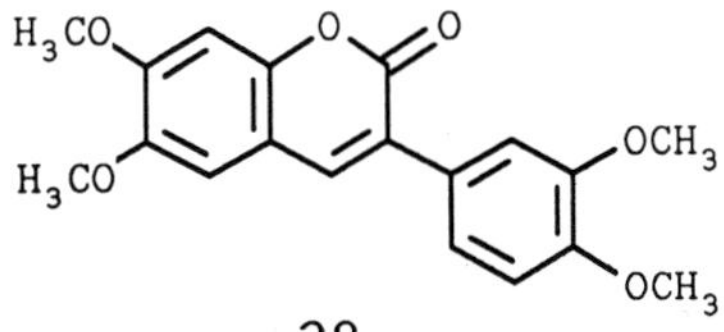

38

39

40

41

42

43

44

45

46

47

48

49

50

SUBJECT INDEX

The abbreviations (F, T, P) are used to imply that the item referred to appears in Figures (spectra), in Tables or in solutions to Problems.